Compassion Makes a Difference

Discussions by the
Laboratory Animal Refinement & Enrichment Forum
Volume III

Edited by Viktor Reinhardt
ANIMAL WELFARE INSTITUTE

Animal Welfare Institute
900 Pennsylvania Avenue, SE
Washington, DC 20003
awionline.org

Compassion Makes a Difference
Discussions by the Laboratory Animal
Refinement & Enrichment Forum, Volume III

Edited by Viktor Reinhardt

Cover photo: Brittany Randolph | Flickr Creative Commons
Design: Ava Rinehart and Alexandra Alberg
Copy editing: Dave Tilford, Cathy Liss, and
Annie Reinhardt

Copyright © 2013 Animal Welfare Institute
Printed in the United States of America

ISBN 978-0-938414-82-7
LCCN 2013939212

I dedicate this book to the investigator
who does not treat animals as disposable research tools
but as sensitive creatures whose well-being
determines the quality of biomedical research data.

Table of Contents

CATS & DOGS

1 Environmental enrichment for cats
3 Environmental enrichment for dogs
10 Elevated resting surfaces for dogs
12 Playroom for dogs
14 Cleaning dog feeders

PIGS

15 Pair formation of pigs
16 Encouraging rooting behavior in pigs
19 Treats as training tool for pigs
21 Oral dosing of pigs
22 Multiple blood collections from pigs

SHEEP & GOATS

25 Environmental enrichment for sheep
27 Environmental enrichment for goats

COLD-BLOODED ANIMALS

29 Environmental enrichment for cold-blooded animals

RODENTS

31 Environmental enrichment for rats
34 Aggressiveness in single-caged rats
35 Species-appropriate housing for mice
37 Bedding and nesting material for mice
43 Usefulness of the term *enrichment*
44 Preferred nesting location of mice
44 Mouse cage changing

45 Treating UD [ulcerative dermatitis] in mice
45 Foraging enrichment for rodents
46 Shelters for rodents
49 Wheel-running in rodents
52 Rodent enrichment — census
53 Training rodents to cooperate during procedures

RABBITS

57 Environmental enrichment for rabbits
62 Housing female rabbits in pairs
64 Bunny nest
65 Protected social contact housing for male rabbits
67 Acclimating rabbits to humans
71 Oral dosing of rabbits
72 Recognizing pain in rabbits

PRIMATES

73 Access to the arboreal dimension for monkeys
75 Play cages/areas for monkeys
76 Woodchip bedding for indoor-housed macaques
77 Foraging and feeding enrichment for monkeys
91 Gnawing sticks for monkeys
94 Water as enrichment for monkeys
96 Protecting watering system hoses

96 Pair and group formation of monkeys
106 Intermittent pair-housing of macaques
107 Pair-housing macaques of different species
108 Oral dosing of monkeys
115 Preparing monkeys for handling procedures
116 Monkeys cooperating during procedures without formal training
118 Monkey see, monkey do
120 Training monkeys to enter into a transfer box/cage
125 Training macaques to cooperate during blood collection
127 Training macaques to cooperate during sedative injections
128 Training macaques to cooperate during saliva collection
130 Are male macaques more difficult to train than females?
131 Treats as training tool for macaques
132 Training vervets to cooperate during procedures
133 Training and behavioral pathologies in monkeys
135 Making use of quarantine time
137 Behavioral pathologies in macaques
140 Touching non-human primates

▮ MICELLANEOUS

145 Dealing with repetitive locomotion/ movement patterns
147 Addressing social needs of animals who have a bandage
149 Radio sounds in animal rooms
151 Phased lighting in animal rooms
152 Who is in charge of environmental enrichment?
155 Environmental enrichment and data variability
156 Animal welfare and good science
157 Are scientific benefits balanced against costs to research subjects?
159 Naming animals in research laboratories
161 Higher versus lower animals
165 Are animals aware of themselves?
167 Do animals have a sense of humor?
169 Retiring and adopting animals who are no longer needed for research
177 Human-animal trust relationship
180 Dealing with emotional fatigue
183 Communication between animal care staff and investigators
186 Professional satisfaction

References 190
Photo Credits 197
Index 202

Introduction and Acknowledgements

This is the third volume of discussions that took place on the Laboratory Animal Refinement & Enrichment Forum (LAREF). This forum is dedicated to the exchange of personal experiences of refining the conditions under which animals are housed and handled in research laboratories. Interesting questions and relevant answers and comments were selected that were posted by 88 LAREF members between March 2010 and December 2012.

I am grateful to the following animal caregivers, animal technicians, clinical veterinarians and researchers for editing together with me their contributions and giving me permission to include these in this book: Dawn Abney, Jason Allen, Krystle Allison, Genevieve Andrews-Kelly, Paula Austin, Kate Baker, J.B. Barley, Sharon Bauer, Vera Baumans, Paula Bazille, Krista Beck, Lorraine Bell, Kaile Bennett, Louise Buckley, Rebecca Brunelli, Lori Burgess, Holly Carter Anderton, Lynette Chave, Yvette Chen, Cathryn Coke, Michele Cunneen, Ernest Davis, Heidi Denman, Louis DiVincenti, Francine Dolins, Marcie Donnelly, Richard Duff, Michel Emond, Fabio Fante, Anthony Ferraro, Thomas Ferrell, Kelsey Finnie, Alyssa Foulkes, Renee Gainer, Alison Grand, Stefanie Haba Nelsen, Catherine Hagan, Amanda Harsche, Penny Hawkins, Harriet Hoffman, Melissa Ann James, Kenneth Jones, Hannah Kenward, Amy Kerwin, Amy Kilpatrick, Judith Kirchner, Michael Kob, Ann Lablans, Jennifer Lofgren, Shelley Lower, Meagan McCallum, Keely McVeigh, Kristin Mayfield, Darren Minier, Robin Minkel, Amanda Moitoza, Erik Moreau, David Morton, Ali Moore, Zachary Myles, Darlene Potterton, Octavio Augusto França Presgrave, Mary Rambo, Kimberly Rappaport, Jillann Rawlins-O'Connor, Angelika Rehrig, Dave Robertson, Susan Rubino, Jodi Scholz, Polly Schultz, Jacqueline Schwartz-Cohoon, Jürgen Seier, Meagan Shetler, Chris Sherwin, Evelyn Skoumbourdis, Adrian Smith, Carolyn Sterner, Autumn Sorrells, Karena Thek, Lydia Troc, Melissa Truelove, Pascalle Van Loo, Pascal Van Troys, Augusto Vitale, Dhaval Vyas, Richard Weilenmann, Christina Winnicker and Russell Yothers.

I have added in brackets editorial clarifications and supportive references from the published literature. Great thanks are due to Cathy Liss, Dave Tilford and my wife Annie for correcting errors and flaws in the final draft of the text.

To obtain photos from animals in research facilities is—understandably—problematic because such material can easily be misused. Therefore, most of the accompanying photos of this book were obtained from the public domain, especially from Flickr's Creative Commons. I am particularly grateful to Polly Schultz, founder of the OPR Coastal Primate Sanctuary in Longview, Washington. Polly shared numerous photos from the sanctuary's animals to accompany her own comments and those from other LAREF members as they relate to refined housing practices for non-human primates.

Alexandra Alberg and Ava Rinehart prepared the layout and created the design of this book. They added a gentle touch of beauty which makes me very happy. Working with Ava and Alex as a team in a relaxed but committed environment was a real privilege for me. Thank you Ava! Thank you Alex!

LAREF is managed and moderated by Marcie Donnelly, Erik Moreau and Viktor Reinhardt, who reserve the right to accept or reject subscribers. If you want to join the forum, send a message to *viktorawi@yahoo.com* indicating briefly your practical experience with animals in research, your current professional affiliation, and your interests as they pertain to this discussion group.

Viktor Reinhardt
Mt. Shasta, California
May, 2013

CATS & DOGS

Environmental enrichment for cats

How do you help cats deal with chronic boredom when they are kept alone?

The room with our singly housed cats has a window that allows all the cats of the room to look out to the animal care hallway. They love to sit and watch the day's business go by. I'm sure an outdoor window would be even more popular but that isn't feasible.

When we had cats, those in my care always received extra human attention. Some of them wanted to be petted or groomed while others preferred to just watch me doing the chores. Most of them enjoyed it when I played with them, throwing a small ball in the room or dragging a piece of rope with a toy attached to it.

Each of our cages is furnished with a hammock or a comfortable raised resting board which all the cats seem to love, plus they have toys hanging from the cage ceiling that they enjoy batting at from time to time.

It has been my experience that single-caged cats are not easily distracted by any kind of environmental enrichment gadgets, but they all love their raised, comfortable platform from which they can monitor what's going on in their room. Depending on their experience with humans, some cats readily trust me and enjoy

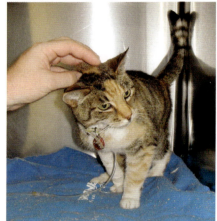

it when I visit and groom them. Others are better left alone because they have difficulties overcoming their mistrust of humans. However, they also enjoy watching from a safe distance what is going on in the room. I believe that the unobtrusive presence of the attending caregiver is the most effective enrichment that can be offered to single-caged cats.

Once they have overcome their distrust of humans, most cats like it when you interact with them in a friendly, cat-appropriate way. Under those circumstances, the attending care personnel can provide high-quality environmental enrichment, especially for single-caged individuals.

I would say that pretty much all of our cats actively solicit human contact. Our cats are handled in a friendly manner by the husbandry, clinical and research personnel several times a day. They are with us for several years, most of them have been here for about seven years; this allows us to establish affectionate and trusting relationships with them. As a result of this, most of our cats are more dog-like in that they initiate playful interactions with you; they would love to have you around all day long. Besides all of the positive handling and interactions with humans, our cats have elevated fleece beds and various toys, and they all get out-of-cage exercise, moist food and catnip once a week as a treat.

I have worked with singly housed cats in a shelter setting. It is my experience that they like it when you interact with them either playfully with a feather teaser or

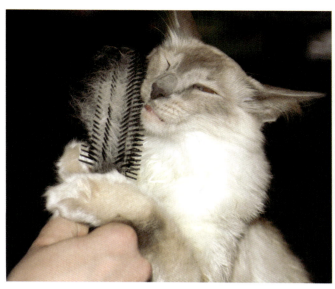

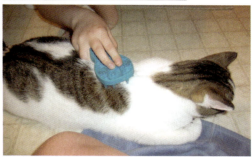

affectionately, by brushing them. It fosters a trust relationship that can really make it a lot easier to handle them if needed. Cats like it when you leave the brush in their cage; they will rub up against it, lick it, push it around, and rub up against it again—until you return and continue the grooming. Depending on their temperament, some cats enjoy interacting with each other on the floor of the room while I clean their cages.

Environmental enrichment for dogs

What kinds of toy are favored by dogs in the research lab setting?

I find that dogs lose interest in toys that are not interactive, do not change their appearance and form and do not provide any social play reward.

Chase-n-catch the ball in the hallway is a game my dogs can't get enough of. They love it when I let the ball bounce back from the wall over and over again until I get tired, not the dogs!

Any toy loses its value for most dogs after a while if a human is not attached to it.

I did a study in a shelter that could be summarized as *puppies play with everything, adults lose interest unless there is food or people involved.*

We haven't had dogs for many years but when we did I had many enrichment devices and would try to rotate them daily. I am very big on recycling things and there is always a money issue in a lot of facilities when it comes to enrichment; so by recycling I could give our dogs more for less. Below is a list of a few such items that I used regularly because they all entertained the dogs quite a bit:
> Paper bags from bedding or rodent chow, stuffed with shredded paper with treats added. The dogs loved to rip these baited bags open.
> Old mouse bottles hung by a piece of rope from their cage door. I would put treats in the bottles and the dogs would have to nose at them or turn them upside down with their mouth to get the treats to fall out.
> Closed cardboard boxes filled with hay or shredded paper and a few treats added.
> Paper towel rolls baited with treats and both ends stuffed with paper towels.

These daily rotated enrichment items have proven to be very attractive for the dogs, but I have to emphasize that the most important enrichment for them is daily play time with one or several compatible canine buddies along with daily, relaxed and friendly interaction time with us humans who care for the dogs.

For hygienic reasons it would be preferable to hang toys for dogs off the floor. The question is: would a dog actually show interest in such a dangling, potentially very attractive toy beyond the initial exploratory sniffing?

I had good success with treat-dispensing toys called Buster Blocks; the dogs were very interested in those gadgets as long as they could get treats from them. Once empty, though, they didn't bother too much with them. While hanging toys do stay cleaner, I think it fair to provide the dogs with toys that they really enjoy and use, even though they lie on the floor and hence may need to be washed more often.

We tried to make the toys more interesting by hanging them off the floor, but noticed very quickly that it didn't make any difference to the dogs: they pretty much ignored the toys whether they were lying on the floor or dangling from the ceiling.

A way of keeping suspended toys attractive is by interconnecting them between different pens or cages; whenever one toy of the chain is pulled down by a dog, the toys of the neighboring dogs are spontaneously swinging, thereby enticing attention and often also active interest.

I don't know about dogs; however, pigs will play with hanging toys much more than toys on the floor.

What could be the reason for the fact that pigs, unlike dogs, are so interested in items that are suspended from the ceiling or hung from the walls of their enclosure? For example, a pig, unlike a dog, will play with a strip of cloth or any other pliable object that is attached to a suspended chain until the object is worn down and needs replacement.

In my experience, dogs like to hold a toy. They usually pin it between their paws or against the floor and have a good old gnaw. Pigs root and seem to enjoy movable toys. They will sample-taste a toy, then root it around and sample it again. This can keep them busy for quite some time.

I am no pig expert, but I am wondering whether this might be due to the pigs' feeding habits. They are omnivores and eat berries, nuts, fruit, etc. In the wild or free range, pigs may sometimes pull these from bushes and low tree branches. Perhaps, the hanging enrichments you mention simulate this way of feeding. A dog however, is a carnivore and, in the wild is only likely to find his food on the ground, so hanging toys are probably not at all interesting for him [unless these are baited with favored food].

We had Kong toys hang from a short chain from the top edge of the cage, about at nose-height of the dogs, who would sniff them when first placed, but otherwise they ignored them. We also tried rope-toys attached with a bungee cord to a cage wall, thinking the dogs could kind of play tug of war with themselves. A few dogs managed to snap the bungee: rope

CATS & DOGS   5

falls to floor, dog plays with the rope a little bit and then ignores it—until a person comes into the room and gets hold of rope; the dog is now showing intense interest in playing tug of war with the person.

Group-housing offers species-appropriate enrichment for dogs; pair-housing is more practicable because you can match compatible personalities and control possible aggression at feeding time.

Based on your own experience, what kind of practicable environmental enrichment do you recommend for dogs in the research lab setting?

At our facility, all dogs are housed in compatible pairs or trios. It is so wonderful to see them romp and snuggle together. I think housing with one or several companions is a great means of enrichment for dogs. We separate companions only temporarily during feeding times, just to make sure that nobody gets too greedy and starts trying to monopolize the food. We have wonderful vet techs who are responsible for ensuring that the canine pairs or triples are compatible. We have not had serious problems with incompatibility so far.

In addition to the social-housing, our vet techs have human socialization time with the beagles, which consists of several dogs and techs playing in a large area.

It is my experience that humans can provide excellent environmental enrichment for dogs in research labs; it serves not only the dogs and the caretakers but it is of great value also for research methodology by minimizing fear/anxiety during handling procedures.

If you are in charge of dogs, do you find the time to interact with your animals in a relaxed, playful way on a regular basis?

Yes, regularly interacting with our dogs is acknowledged as part of our enrichment program not only at the facility where I am working but at five other sites of our company.

We too have regular relaxed human interaction with our approximately 400 dogs incorporated in our enrichment program. The dogs are pair-housed; they are regularly released so that they can run up and down their rooms, but also sometimes in the long hallways when the rooms are sanitized. We have great animal care staff and depend on them to be the primary people interacting with the dogs; however, we also have a program where PIs [principal investigators] and chemists can come over and play with the dogs in a designated playroom.

Back when I cared for rabbits, I shared some duties with a colleague who was in charge of 18 mixed hounds. She had established several compatible groups and we played with them on a regular basis. Each dog got our personal attention through playing, grooming and gentle talking. We also had a fenced-in area outside the dog runs where we would take the dogs out to play whenever the weather was fine. We had pools for them in the summer, and special outside toys. We often spent our lunch breaks in the company of the dogs. I was very proud to be involved with that program.

When I worked with six dogs, we too made time for daily human interaction—very important in my opinion for both the animals and the staff. Each dog had individual time for grooming, training and snuggling, as well as supervised play sessions in compatible groups in the holding room. We also trained the dogs to walk on leashes so that we could walk them around inside the facility. They also learned to accept teeth brushing, which was a frequent procedure for them as part of the dental study they were assigned to.

Many years ago when we housed dogs, each one got human interaction daily. Unfortunately, we did not have an official socialization program, so we had to give up some lunch time to be with our dogs. Usually, they were housed in pairs, but sometimes that was not possible. When we had to single-house, each dog was let out of his or her cage while I cleaned it. After cleaning I would sit with the dogs individually and interact with them in whatever form they preferred. This could be petting, scratching, playing ball, or grooming. Some breeds like German Short-haired Pointers were not real fussy about being scratched and petted but they sure loved a good game of fetch the ball.

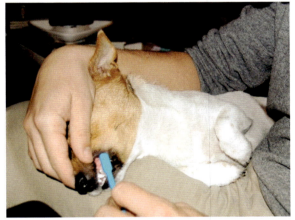

I formulated an Enhanced Canine Enrichment Plan for our dogs; technicians go through a dog behavior refresher course and then are allowed to play with our single-caged dogs in the anterooms whenever there is a lull in their schedules. So far it has been difficult to get a time commitment from enough staff members so that each dog gets direct, friendly human attention on a daily, rather than occasional basis.

Regular relaxed interaction with the human caretaker(s) is so critical in obtaining quality research results that it should be an integral part of any dog enrichment program. To *allow* attending staff to socialize with the dogs in their care is not sufficient; socializing with the dogs has to be part of the technicians' job description so that they can really commit themselves to this important responsibility during their regular, paid work hours.

It seems to me—but I may be wrong—that facility administrators, vets and investigators are more willing to have the social needs of dogs addressed than of macaques, both in terms of socialization with conspecifics and socialization with attending personnel. If this is correct, what could be the reason for this bias for dogs?

I think you are correct. Dogs are truly domesticated animals; usually you can walk into their kennel and interact with them in a playful or affectionate manner without fear of being attacked. Many people have a dog as pet in their home, so they have some basic knowledge of a dog's emotional, behavioral and physical needs; often they have developed an affectionate bond with their dog, who has become a cherished companion.

With macaques, you cannot simply open their cage and interact with them without risk of being attacked. Macaques are undomesticated animals who typically mistrust humans as potential predators, and hence often show aggressive self-defensive reactions toward them. Most people have very little knowledge about the behavioral and emotional needs of macaques, so they would probably be much less concerned about species-appropriate housing conditions of macaques than of dogs.

I would say that the bias for dogs stems from the image that we have of them as companions. Thousands of years of co-evolution have affected our emotional relationship with dogs. This relationship is based on friendly physical interaction that strengthens attraction and empathy. You can hug a dog and he will lick your face; thus, an affectionate bond is formed. Once a positive physical relationship is established,

you are more likely to attend to the welfare of that animal. With macaques in the captive environment, positive physical interactions are very restricted; therefore, you are less likely to develop affectionate, close relationships with individual animals. Personnel may be attracted to those individuals who behave in a friendly manner, for example lip smack or present for grooming. The assertive, cage banging macaque who threatens and grabs at everyone who passes by is associated with negativity and, hence, not likely a recipient of caring attention. The same, by the way, is true for dogs who are overly assertive and threaten anybody who approaches them. There are probably not many caretakers who are patient and compassionate enough to work with such animals, so their welfare needs are more likely to get a bit ignored.

If animal care staff, administrators, veterinarians and investigators feel a closer connection with dogs versus macaques and hence are more inclined to address their needs for well-being, why have conspicuously more scientific articles been published on environmental and social enrichment for macaques than for dogs? The two Refinement and Enrichment Databases as of May 1, 2012, listed 972 entries dealing with macaques versus only 144 entries dealing with dogs. It is my understanding that the number of dogs used in research is comparable with the number of macaques used in research. Is it perhaps more interesting or fancy to do environmental enrichment studies with primates than with canines?

I think that the discrepancy in the quantity of articles stems from many sources, one of which may be the differential care given to dogs versus macaques. If there is a perceived bias or favoritism towards dogs, then the care given to them is perhaps better than that given to macaques; this could result in relatively few behavioral pathologies in dogs kept in research labs.

The literature on environmental enrichment for macaques may be vaster than for dogs because macaques develop many more behavioral pathologies in the research lab than dogs, hence more research on alleviating those problems is conducted and the findings are published in many scientific articles.

The more complex social structure of macaques—as compared to dogs—makes socializing them in the research lab setting a more challenging proposition. This could be one reason for the larger number of scientific articles pertaining to species-appropriate social housing for macaques than for dogs. I don't think that it's because primates are more interesting or fancy: I think that our understanding of their behavior and social system is still evolving, and as such garners more scientific research.

Elevated resting surfaces for dogs

U.S. Animal Welfare Regulations for dogs and cats stipulate that "each primary enclosure housing cats MUST [emphasis added] contain a resting surface .... The resting surface must be elevated" (United States Department of Agriculture, 1995).

This requirement is very laudable, but it is surprising that it is restricted to cats only. Do not dogs also benefit from an elevated, dry, lookout resting surface (platform/board/shelf/bed)?

We have elevated bed-boards in our facility. The first thing we noticed when we moved into this facility several years ago was the dramatic reduction in both the duration and the volume of barking. We attributed this to all the dogs of the room being able to see immediately and simultaneously the cause of the initial barking, i.e., personnel entering their room. In more traditional facilities, only those dogs closest to the door start barking when a person enters the room, and this then triggers a kind of chain-reaction from all the rest of the dogs who are unable to see the cause of the excitement and hence will continue barking until they have all seen the person who has entered their room.

The dogs also appear to prefer elevated bed-boards, presumably because these give them an increased visual range and help them to establish and maintain a social hierarchy within the room.

I have in the past used Kuranda beds; they were a huge success! Often we would see the dogs sleeping on them; they were so comfy that our presence didn't really disturb the dogs. They chewed on the edges, though, and sometimes would dig through the beds. As a result, the beds got a bit worn out. But, like any enrichment, the destruction showed that the dogs were definitely using them.

In the facility where I currently work, we use resting boards. I have observed dogs hiding toys under them, and then digging them out of the hiding space. Sometimes they seem to use the platforms to gain height over their cage mate(s).

I think all dogs in research labs should have access to an elevated resting surface. We have two groups of four beagles. Their pens are furnished with several tables. The dogs sleep on/under them and use them as a lookout platform. Jumping on them provides an exercise opportunity. One problem is that the males like to mark the table legs, so we are intending to replace the tables with floating shelves.

I like offering our dogs some means of getting off the floor. Right now, we just use commercially available dog cots. The problem we are having at the moment is that the type we purchased does not hold up well to rambunctious pups. Next time I will probably spend the extra cash and get the ones constructed of sturdier materials. I have found that all the dogs use them and prefer the comfort of the slightly raised bed to the floor.

I've worked at two facilities that had indoor runs for dog housing. One facility had a platform at the back of the cage, the other had no platform. While there was no obvious sign that the dogs without the platform felt they were lacking, I'd say raised platforms should be mandatory. The dogs who had platforms, all used them whether for sleeping, as a way to see more of the room, or to hide under when they were stressed or fearful. Had only a few of the dogs used their platforms I might say "make them optional," but seeing an entire room of dogs using them indicates to me that that the dogs have a need for a raised resting surface; so it should be a standard item of furniture of their living quarters.

There can be difficulties providing platforms if you have escape artists who

climb. For those dogs you need to have a top on the kennel, and I for one constantly managed to hit my head in those runs.

We regularly provide raised resting surfaces for a variety of animals in research labs—including cats, rabbits, rats and non-human primates—and I see no reason why we shouldn't be consistent in providing those for dogs as well.

The dogs I have worked with almost always used their raised platforms to sleep or rest on, hide under, or jump up and down from. We had random-source dogs of various breeds and I can't think of one dog who didn't make regular use of the platform. I firmly believe that dogs need an elevated resting area to feel at ease in their living quarters.

We had indoor runs bedded with shavings that were scooped and replaced daily. When it came to rinsing the runs, the dogs were able to get away from the water hose by retreating onto the raised platform, where they would wait until the cleaning was done.

I think access to elevated sites is very important for confined dogs. Our dog runs have a bench at the back with a sturdy ramp attached to it. The bench protrudes about a foot out from the back wall and spans the width of the cage. Our dogs run, jump and sleep on it, and use it for visual and tactile stimulation with their neighbors.

Playroom for dogs

We intend to establish a playroom for our dogs but have no experience. Does your facility have a designated playroom? What is the typical group size and group composition of dogs that get access to a playroom? What kind of enrichment do you offer the dogs in the playroom?

In my first job in research—more than 15 years ago—I had the privilege of working in a dog colony. We had a dedicated playroom with a tile floor and spray hose for easy cleaning. It also had a large storage bin on wheels for several—actually lots—of toys/enrichment devices; the bin could easily be moved about. There were two holding rooms adjacent to this playroom.

The dogs were brought out, six at a time, from the holding rooms into the playroom for one hour each day. The PIs didn't mind about the playroom as long as I was the one moving the dogs back and forth. I can tell you, moving six hyper-overweight beagles from their cages

in one holding room into the playroom was more fun to watch than to actually do. And the best enrichment device? Me dressed head to toe in a Tyvek suit!

The dogs really just wanted the interaction with me more than any of the toys. These were all intact males, and fortunately I never had to deal with any serious fights. A few growls at first, and I noted who didn't play well with whom and arranged the group composition accordingly. I loved to work with these dogs and would do it again in a heartbeat, but with two modifications:

1. Have an anteroom to the playroom. Opening the door of the playroom in order to let new dogs in, or take dogs out was the biggest challenge. It only took one strong dog to pull the door open and then I had six dogs running in the halls. A transfer tunnel between the holding room and the playroom, with two guillotine doors in both rooms would be ideal.
2. I would also think about a different flooring option. The tile floors got very slippery during the play sessions, and walking on them, wearing shoe covers, with jumping dogs was not always without risk.

Did I mention how me wearing a Tyvek suit was the best enrichment for the dogs?

We do not have a playroom for our dogs. Instead we have gated-off hallways in the building and let dogs out into these hallways while their rooms are being cleaned. Most of the time, various staff interact with the dogs. We have a variety of toys for the dogs to play with, but mostly the dogs simply enjoy being with people; they get to interact with anyone who walks by the gates. Due to the nature of the studies that the dogs are involved with, some of them cannot be let into the play area with other dogs, but all the others come out in groups. I do like the idea of a dedicated playroom. We just do not have the space at the moment.

CATS & DOGS 13

Where I used to work we had only a couple of dog rooms. We would let all 6–8 dogs from a room out while we placed toys on the floor and hosed cages; when the hosing was done, we spent 20–30 minutes playing and interacting with the animals.

The facility where I currently work has a designated playroom furnished with kid's slides, tunnels, and lots of chewable toys. Vet staff keep the dogs company, petting them, playing with them and enticing them to play with the toys. From an outsider's perspective, the dogs appear to be having a BLAST!

Cleaning dog feeders

I am interested in finding out how often those of you who house canines in your facility change out the feeders. Also, have you experienced any mold in the dog food as a result of not changing them often enough? And if the food is getting moldy, do you believe it is from moisture already present in the food or just water getting in the feeder?

It's been quite a long time since I worked with dogs but I found that metal dog feeders had a tendency to stay wet. Frequently, bits of food got stuck in the corners and would get moldy in a short period of time—on the order of 24 to 48 hours. Initially those feeders were sanitized once a week, but after seeing how frequently there were bits of moldy feed, I switched to sanitizing them daily.

My preference would be to not use the style of feeder that attaches to the run but use bowls instead because these are much easier to clean manually.

I agree that bowls are much easier to clean than the rectangular feeders. Food gets stuck very easily in the corners of traditional feeders, where it gets wet and can mold fairly quickly. We actually removed all our metal feeders and replaced them with plastic bowls. The bowls are sprayed out every day, and sanitized every two weeks when we feed dry food, and usually every week when wet food is added to the dry ration.

PIGS

Pair formation of pigs

What's the best way to pair pigs with each other? Should potential partners be familiarized before they are placed together, much like macaques?

We have so-called enrichment panels in between cages, allowing neighboring pigs to see, touch and smell each other. I like to pair pigs who have been next to each other for at least a few days. It's very rare that we don't see some fighting when the partners are then introduced to each other in the same enclosure. They need to work out who is dominant; once they have done that, they calm down. I typically watch them over the course of the day. If the fighting gets too rough I separate them, otherwise I let them stay together. The problem is that the pigs are randomized into studies, which means pairs are often split because partners end up in different studies. I keep track of who has already been paired with whom; so if the opportunity arises these animals will be allowed to live together again.

Have you seen much stress when pairing the pigs with different buddies? I read an article saying that it took about two weeks for elevated values of physiological stress parameters to become normal again.

I haven't measured any parameters, but I can say that the two pigs usually are stressed the first one or two days while they are getting used to each other; after that they seem to be fine, sleeping huddled together, showing normal behavior, normal attitude, and normal appetite. In the beginning, I will give both of them treats and normal food at the same time

so that they get used to sharing food rather than fighting over it.

I think the tactile familiarization time before the pairing helps, but it doesn't completely eliminate the initial fighting and stress.

When we take delivery of new farm-reared pigs, we always sedate the animals so that they fall asleep; we then pair them and place the partners in such a way that they are touching each other [thereby transmitting each other's personal scent]. When they finally wake up together, they seem to be quite happy with each other. We give them enrichment in the form of deep straw, cardboard boxes and portable objects, along with lots of human contact to keep them entertained; we rarely encounter problems.

At my old workplace, our technicians applied lavender oil on the backs of new minipig arrivals and only then paired the pigs up. This worked quite well; there were no aggression-related problems.

Lavender essential oil is known for having calming properties so it may decrease aggression.

Species-appropriate rearing conditions seem to help pigs sort out dominance-subordinance relationships without much distressing aggression. O'Connell & Bettie (1999) showed that pigs from an enriched rearing environment—with provision of extra space and substrates for rooting—fought significantly less with unfamiliar animals than those reared in a barren environment: "Enrichment appeared to facilitate the development of social skills which resulted in body weight, rather than aggression, determining dominance."

Encouraging rooting behavior in pigs

We are re-evaluating the use of materials to encourage foraging and rooting behaviors in pigs and I'd like to get a sense of what other places are doing.

How many of you are hiding treats in hay, straw or shredded paper? We've been using hay and straw and the question has been raised about switching to shredded paper. Has anyone observed pigs actually consuming the paper? What type of paper have you used?

Has anyone tried going green by shredding the packaging used for bedding and food? We do give our pigs the bedding bags on the days they get their bedding changed; this appears to be a great form of enrichment—while it lasts!

When housing the pigs in a raised system where treats and goodies can fall through the grates to the floor, how do you encourage rooting behavior?

Our pigs are housed on solid floors with separate sawdust-substrate bed trays that have raised sides of about 4 inches [10 cm] and are provisioned with fresh straw daily. We hide food treats in the straw, but our pigs engage in rooting quite happily even when given fresh straw without any incentive. Straw tends to repel water to some extent and although it gets wet, it doesn't stick—a big advantage over paper-based materials! We have also used shredded office paper with our pigs; they are just fine with it and there are no signs of them eating the paper. BUT when wet, paper can turn quickly into a nasty soggy mess sticking to the floor, bars and any other parts of the pen. We encountered the same problem with soft paper, which even stuck to the pigs' skin. The pigs benefited from this as they enjoyed having the paper brushed off their bodies. The big drawback with shredding paper is that it's time consuming and mind numbing; we have never found a shredder that you could just walk away from, leaving the machine to do its job without human assistance.

We don't use newspaper, as the pigs go gray when the ink comes off; this compromised our routine health checks as we use the color of the skin as an indicator of the individual pig's health status.

I use shredded office paper to hide treats in and haven't had a problem with the pigs ingesting the paper. The pigs do chew it from time to time but, as far as I can see, they spit the chewed paper out without eating it.

I fill cardboard boxes with either hay or shredded paper, then close the boxes and let the pigs rip them open to get to the treats. I also take old water bottles for mice, punch a hole near the top rim, and then tie them together on ropes and fill them with treats. I suspend them from the wire at the front of their pen, so the pigs have to root them upside down to retrieve the treats. It doesn't take them long to figure out what they have to do to get the goodies. I also create what I call a rooting basket. I take a large rubber feeder—often used on pig farms—and put four stainless steel bowls in it upside down with treats hidden under each bowl. I then fill the feeder with shavings, shredded paper or hay. The pigs really enjoy rooting around, righting the bowls so they can get to the treats.

PIGS   17

A basketball [or a pumpkin, as shown above] provides great enrichment for pigs. They love to root the ball around and try to bite it and pick it up; often two pigs [or a sheep and a pig] will join in the fun and make a game of it.

A long piece of rope or cloth with knots tied in it, or old rubber sipper tube stoppers threaded on it will be rooted and shaken around by the pigs and also used to have a tug of war with a buddy. The rope is easily cleaned in cage wash and the cloth simply replaced when dirty.

I have heard of people making rooting boxes, using a wooden box with high sides that is partly filled with a mixture of rocks and treats. The pigs have to root the rocks around to get the treats.

We have had good success with the J-shaped feeders that have a flap on top, covering the food. The pigs have to root them open to get at the pellets.

We house our swine on raised flooring. To encourage the natural behavior of rooting, we purchased a heavy stainless steel chain and placed that in the pigs' pens, three links for the small Göttinger pigs and six links for the large swine.

The pigs appear to really love the chains, as the techs hate them. With the pretty much constant pushing and banging of heavy metal there is so much noise in the room that hearing protection is required. The large pigs are literally throwing their chain across the room, making it quite an adventure for the staff to be around.

At my previous facility we also kept the pigs on raised floors, but were able to turn an area with solid flooring into a playroom where we hid treats in pine shavings. The pigs would ingest some of the shavings, but it never interfered with their digestion or food consumption. Single pigs or groups of pigs rotated through this playroom such that each animal could use it at least once a week. The animals were left to root and play for two hours, and then returned to their raised home pens.

For inside their pens, we used heavy perforated balls that we could place treats in. The balls were suspended on chains and the pigs had to push them up and down with their nose to shake the treats so that they fall out. The hard part was finding treats small enough to fit inside the balls but big enough to keep them from falling through the flooring.

I put certified enrichment into recycled, large plastic tubes that have Marshmallow Fluff in them and are closed at one end, let them sit in the freezer overnight, and give them to the pigs the next day. I recently watched a couple of them trying to keep their tubes in an upright position and stick their snouts inside. They would try and hold the tube, but it would slip and slide to the other side of the cage only to tip over. They repeated this over and over again. What a riot! Finally one of the huge farm pigs got the idea to corner his tube—and he got what he wanted! It was really funny to watch them; they were all so engrossed in this challenging rooting game.

Treats as training tool for pigs

Is anyone on the forum willing to share experiences concerning training and/or enrichment for pigs? What have you trained them to do? I know pigs are intelligent, so are they easy to train? What types of enrichment, treats, etc. do pigs like?

One of our labs is trying to clicker train their pigs, starting with their current solo male pig. He loves his daily rations but refuses to work for—or even take—any treats. They've tried marshmallows, fruit, veggies, peanut butter—no go.

I also had pigs who would not take treats spontaneously. To get them to cooperate, I smeared peanut butter with a finger on the roof of their mouths. I had to do this exercise only a few times, and most pigs would eagerly lick the peanut butter after they got the taste for it.

I also did this with strawberry jam, which they really enjoyed, and fed them cereal as a treat. A few days after their arrival I would place a pile of hay in their pen and scatter cereal throughout it, and then let them root for the treats in their own time. Once they tasted them, they willingly took them from my hand.

I have worked with several types of swine and have found there are a number of differences when target training Yucatans, farm pigs (Yorkshires), and minipigs for intramuscular injection, restraint in slings, and presentation of body parts. For me, Yucatans are the easiest to train—and some of the greatest swine you will ever meet!!—and minipigs are the most difficult due to their high-strung nature. The farm pigs I've come across varied quite a bit in their personalities; some were easy and others were difficult to work with.

As for enrichment, I've found that jelly beans, PRIMA-Treats, apples, monkey biscuits and dog biscuits are some of the most desired food items of pigs. They adore anything they can root along the floor—such as large plastic balls, large cardboard tubes and boxes, and large Kongs filled with Marshmallow Fluff. If you are able to give

your pigs hay, they will go ga-ga. They'll flip it around and happily oink while rooting through it for goodies. Another great entertainment is bobbing for floating apples: we cut apples into appropriate sizes for the type of swine, and drop them into the water bowls. The pigs have a BLAST rooting through the water for the apples!

Pigs love apples and there is no reason to believe that they wouldn't do tricks in order to get them.

I have used peanut butter or jam for training with great success! Belly scratching and rump scratching are just as attractive rewards as treats for pigs who are socialized with humans.

My experience with target training has shown me over and over again that pigs are very smart and learn quite fast. Prang, apple juice, and Karo syrup work wonders as rewards for cooperation!

Yes, pigs really love the sweet Prang drink. I had one over 300 kg boar work with me during jugular bleed training sessions. Unfortunately, he was scheduled to leave us before we could get him successfully trained, but is was very apparent that he would do back-flips for the Prang that I delivered for him with a squeeze-bottle during the training sessions.

We found that our pigs like Ensure—some strawberry flavor, some vanilla, others chocolate flavor. Granted, these animals were post-op and we were trying to get them to eat, but maybe even a non-post-op pig would like it.

Depending upon possible calorie restrictions, I've used Karo syrup to give oral meds to pigs—just be warned, it can be very messy. I find that pigs also like Jell-O, usually the lime or strawberry flavors.

It is my experience that pigs like dried banana chips and readily work for them.

In addition to all the other suggestions, I use whipped cream to make pigs walk onto scales or into the operating room. After I have sprinkled a whipped-cream trail, the pigs follow it to the destination without much hesitation.

Oral dosing of pigs

Can anyone on the forum please share experiences with dosing pigs via pills or tablets? I have already played with the idea of contacting Bio-Serv to have the test compound made into a palatable treat, but I was not sure what flavor would be most appealing to a pig. I was thinking of putting the pills in bananas as they have such a strong flavor and my pig at home relishes them.

I have good luck with Nutri-Grain bars, Fig Newtons and Fruit Roll-Ups. The pigs love them! I would suggest that you introduce these snacks *before* you put pills in them so that the pigs know ahead of time that they receive tasty treats!

We used to stuff pills into apples. It worked fairly well. We also used Marshmallow Fluff—the marshmallow topping that comes in a jar—to coat the pills.

Yes, pigs love apples. The taste of apples seems to cover up the taste of most drugs/medicines. We cut the apples into two pieces in which we hide the pills and then offer the pieces by hand. It has been my experience that minipigs accept the baited apple pieces without any ado.

Sometimes we simply mix the pills with the regular chow ration. That usually works well if you can take the time to check that the pigs haven't spit the pills out. Marshmallows are great because the pills stick firmly inside of

Multiple blood collections from pigs

We have a new study coming up in which it will be necessary to collect multiple blood samples via indwelling jugular vein catheters from Yorkshire pigs (60 kg or more). Blood samples will be taken from each pig six times an hour, for eight straight hours, three days a week for two weeks. The researcher—and the rest of us—would like to do this with minimal restraint and stress for the individual pig, but we are limited in space and equipment. I have already volunteered to target-train the animals, but we're going to have to find a way to confine them during the eight-hour period as all of our pens are raised and too narrow to allow more than one person in the pen along with a pig. We have considered building a floor pen of sorts, but already know that if piggy wants out he/she is going to get out.

Can anybody on the forum please share experiences pertaining to our project?

Wow that's a lot of time points! I assume these indwelling catheters will be surgically implanted and connected to subcutaneous VAPs [venous access ports]. We had some pigs with VAPs a while back; we placed an angled Huber needle with approximately 6 inches [15.2 cm] of tubing with an injection cap at the end the night prior to the study and wrapped—not too tight!—about 3 inches [7.6 cm] wide vet wrap around the pig in a crisscross pattern. This arrangement made it possible that one person could enter the pen, collect the blood with a blunt-tip needle, and flush and lock the VAP catheter. We restrained the pigs in a sling for Huber needle

them, but first you have to get the pigs familiar with the marshmallows and make sure that they actually consume them.

If we know in advance that the pigs are scheduled for pill treatment we usually will first get the animals used to eating a special, canned dog food consisting mostly of corn. The pills can easily be hidden in little food balls.

We had over 100 pigs to be treated with a pill, the taste of which they didn't like at all. Many treats were tried to hide the pill, but the pigs stubbornly refused to cooperate. Finally, frozen cookie dough was tried. The pigs really liked it and swallowed the test pill while eating the tasty dough.

22

placement only. During the study the animals stayed in their own familiar pen.

It has been my experience that with gentle firmness and favored food rewards—especially Prang—it is pretty easy for a person to gain the trust of a pig so that frequent blood samples can be collected for the study and the catheter flushed as needed.

To sustain the cooperation of the pigs for eight hours will probably be a challenge. This may be quite a stretch of time for any beast, especially a pig! A rotation of different attractive enrichment devices and lots of human attention would definitively be indicated to help the pig cope with the situation. I would keep a large crate on wheels close by, just in case!

Thanks for all the suggestions!

We went hunting in our back storage area and unearthed an accordion pen as well and a mobile, relatively spacious transport crate that opens from the top—looks like a farrowing pen on wheels. Finding out from you that with some training the pigs will probably hold still for the whole thing is thrilling! I assumed they would probably run out of patience with me after a while. I've done multiple sampling, but never at this frequency.

I think the main issue is that I'm the only person at my facility who has ever target-trained swine and/or done long-term multiple blood sampling on anything other than an anesthetized animal. So I'm having a hard time convincing folks that, with a little patience, piggy will most likely cooperate.

You should have seen the looks when I presented the Prang idea during our meeting this morning! Thankfully there are no dietary

restrictions, so I can use anything I like to appease and reward the pig. Thus, I'm now getting a standard treat arsenal together to do preference tests well before the study gets underway. I figure, if the poor beast has to cooperate for eight hours straight, then she or he should be ultra happy; the investigator totally agrees.

The researcher has never worked with swine before, but he's one of my favorite investigators here—his care of the creatures is outstanding and he's adopted out more dogs following termination of studies than anyone else here—so I am willing to work with him on this. Luckily I was able to convince him to get the pigs a couple of weeks in advance so that I can acclimate them and target-train them prior to the placement of the catheters and the insulin pumps.

The indwelling catheters will be surgically implanted; they are very similar to a VAP. It's a special infusion pump that the investigator helped invent for human subjects several years ago. He will be doing the surgery, so I have every confidence that all we'll have to do is to train the guys (or gals) to hold still, so we can check for patency of the blood vessel and take samples.

I keep in the back of my mind that patience on the part of the pig may run thin after a while, so I am planning on rotating a plethora of different goodies throughout the day for reward and making sure that the investigative team will use other rewards—such as a good ear or snout scratch—following each blood sample collection. Having trained monkeys to perform for hours on end, I know that the critter will let you know when he or she doesn't want to play anymore. Thus, my plan is to try to read the animal as best as possible and keep the rolling crate or other more spacious holding devices near, just in case.

    The good news is that this investigator is open to any ideas to make this study as easy as possible for the pigs. So my plan is to somewhat spoil them to pieces. Even though most of my experience has been with Yucatans, I'm hopeful that it will help me when working with the Yorkshires. The few Yorkshires I have worked with have been sweet, so I'm really looking forward to getting started. Swine are a favorite of mine and it's been a while since I've been able to give a good snout rub.

SHEEP & GOATS

Environmental enrichment for sheep

We would like to expand our sheep enrichment program. Our drain system works such that the loose hay clogs the drains when the husbandry technician sprays down the runs. Any ideas for a way to give our sheep something to do other than nibbling hay? They don't really care for Jolly Balls and the hay cubes we have are really hit and miss. I was thinking of mirrors, as we have Plexiglas mirrors for the primates, but our head vet heard somewhere that sheep are fearful of mirror reflections. I actually can't find any published research supporting this, so I am wondering if anyone has had any experiences with mirrors as enrichment for sheep.

Have you already tried some mesh-screening over the drains to prevent the hay from entering the drain? It's still going to require the techs to pick up the hay from the screen periodically because it'll essentially clog the screen while they are spraying the runs, but it allows you to continue using the hay, and other types of foraging/browse material.

The provision of hay or straw is pretty much the only source of positive distraction individually penned sheep have. Investing a little bit of extra time in making it practicable to offer them fresh hay or straw on a regular basis is the least we can do for them.

I used to work with sheep who, when they were kept singly, had access to a treadmill.

We had a mirror—polished stainless steel sheet—mounted at the end of the track so the sheep on the treadmill thought she was walking toward another sheep. Not quite sure how it would work for free-walking sheep; might be worth taking precautions in case the sheep wants to attack the reflection in the mirror.

We routinely use mirrors for sheep when we must singly house them. It seems to me that when watching the mirror reflection they do see a pen mate; they spend most of their time nose-to-nose with the mirror, which probably has a comforting effect on them. I would definitely recommend using mirrors if you have to house sheep singly.

We've used mirrors when we had solitary sheep in the unit—usually at the end of a trial. Based on my own observations, I have no doubt that the mirror reflection calms a lone sheep and offers social comfort. Typically, lone sheep stop calling and being agitated when you put a large mirror in front of them.

We used sheets of polished stainless steel or full length ordinary mirrors.

We were amazed at the calming effect of mirrors when we got new sheep. The animals were quite timid and disturbed after being unloaded from the delivery truck and it took a lot of coaxing to have them walk down our animal housing hallway. But as they turned the corner into their housing room, they saw the mirrors and RAN to stand as close to the mirrors as they could. Sheep really do find comfort in numbers even if these are mere reflections of themselves!

Published research indicates that their own mirror reflection buffers stress rather than induces fear in sheep.

    McLean & Swanson (2004) observed that mounting large mirrors on the sidewall of isolation units has a calming effect on sheep: "Vocalization stops completely and the sheep remains completely calm. It seeks out its own mirrored image, stands close and occasionally nudges at its mirrored partner. Consumption of food and water remains unchanged and the risk of injury is eliminated, as the sheep no longer tries to jump or escape the enclosure."

Parrott et al. (1988) found that the presence of mirror panels markedly reduces endocrinological (cortisol and prolactin) stress responses in single-housed sheep. Da Costa et al. (2004) did not use mirrors but pictures of sheep faces and report that the sight of such pictures significantly reduces behavioral (activity and protest vocalizations), autonomic (heart rate) and endocrine (cortisol and adrenaline) stress responses when sheep experience social isolation.

I would not categorize mirrors as enrichment as the sheep can't interact with them in any way. Personally I think the best form of enrichment is another sheep; if this is not possible, large amounts of bulky forage provides species-appropriate enrichment which seems to keep sheep engaged throughout the day. Our sheep are on nutrition trials, so we give them wheat straw, which has almost zero nutritive value. Wheat straw is also handy if you are concerned about your sheep gaining too much weight.

Environmental enrichment for goats

One of my colleagues has two 2-month-old Boer goats living in her garage during the cold winter months. She is looking for ways to entertain these two kids so they stop destroying the garage. Can anybody share experiences with goats and offer some advice?

Goats like things hidden in cardboard boxes; they chew them all up to get to the hidden item. Hay with food scattered in it keeps them quite busy.

Not sure you can guarantee these two kids will not eat the garage.

My director ADORES goats, and happened to be near my desk when I opened my email. So, I got her two cents. Her first comment was browsing, browsing, browsing! She recommends putting long hay strands onto the floor for the goats to chew through. Apparently, hiding hay in cardboard boxes

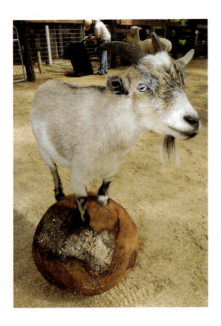

She didn't think balls would be too much fun for these goat kids, as their favorite activity is chewing. [As shown at left, they may not push the balls but use them to exercise their amazing balancing skills.]

I used to work with a large herd of British Saanen goats; they were also winter-housed, as our winters here are very wet and cold and goats are not particularly waterproof. They loved browsing and nibbling/chewing any shrubby or wooden material. Goats are a bit like pigs in that they enjoy chains to play with. Anything that they can destroy will help keep them occupied.

entertains them quite a bit. She said to get mesh boxes or baskets to fill with the hay. Then hang the boxes/baskets so they become sort of a puzzle: make it so the goats will have to bang their heads against the basket or box in order to shake hay out of its container.

Goats are very cute, but kind of professional chewers. Whatever is in their reach—and they are also expert climbers!—is at risk of being destroyed. Any kind of so-called environmental enrichment will probably not hinder these two goat kids to *finish* the garage in the course of the winter.

COLD-BLOODED ANIMALS

Environmental enrichment for cold-blooded animals

U.S. legal and regulatory stipulations pertaining to the welfare of animals used for research and education cover some warm-blooded animals but disregard cold-blooded animals completely (United States Department of Agriculture, 2002 & 2005). Since fish, amphibians and reptiles are not within the scope of the Animal Welfare Act and its regulations, how are they being kept in the research laboratory?

Our African clawed frogs (*Xenopus laevis*) are provided with PVC [polyvinyl chloride] pipes for shelter/hiding, as well as faux floating foliage for additional cover. We've seen a large decrease in startle responses when the tanks are approached or manipulated, as well as fewer injuries from tank mates swimming into each other during such responses. Bullfrogs (*Ranidae*) have access to raised, dry platforms; the tanks are high enough to provide them sufficient space for species-typical jumping. Zebrafish are provided brine shrimp as part of their diet so that they have natural feeding opportunities.

Environmental enrichment for our frogs also consists of PVC-tube shelters, ramps, AstroTurf and fake lily pads. Our fish tanks are not provisioned with any extra furniture but the animals are all socially housed.

It has been documented that frogs prefer enriched living quarters to a barren tank and that access to shelter and refuge-providing structures decreases their startle response and mortality rate while enhancing their reproductive performance and social well-being (Hilken et al., 1994; Hedge et al., 2002; Brown & Nixon, 2004; Torreilles & Green, 2007; Harr et al., 2008; Archard, 2012).

Our frogs also are offered enrichment; the *Xenopus* get PVC shelters and plastic lily pads while the *Rana* have access to a dry platform. The zebrafish are all socially housed and get live food but we do not explicitly enrich

their tanks. We also have a few reptiles at our facility: young Nile crocodiles and tentacled snakes. The crocs are socially housed in tanks with warm, dry resting areas; we feed them live food, i.e., goldfish and crickets. The tentacled snakes live in groups in tanks that are furnished with natural and artificial foliage; they also get live goldfish and crickets.

That's great, all you do for your fish, frogs, snakes and crocs. I'm guessing the crocs are small. Wow Nile crocs—scary when they are big!

The crocs are little, about 10 inches [25 cm] at the moment. They are a bit nasty and will charge against the glass tank walls and if they have a chance will snap at you. Glad I don't have to feed them or change their tanks!

It is very encouraging that at some facilities in the U.S. amphibians and reptiles are being considered as animals who—like warm-blooded animals—also deserve relatively species-adequate living quarters. This proactive, ethical attitude is setting a good example; I am sure not only the animals, but also the research conducted with them and the attending care personnel benefit from it.

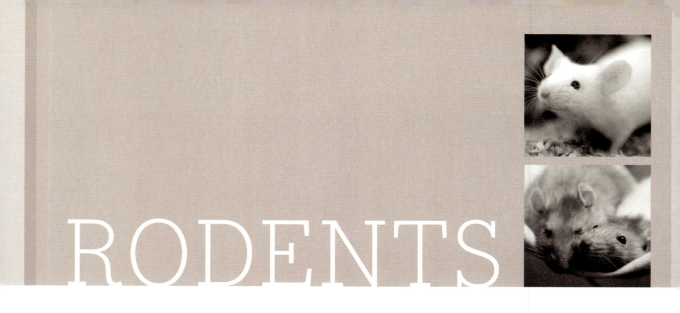

RODENTS

Environmental enrichment for rats

If you had a say and have the experience, how would you furnish the cages of rats in your care in a way that does not compromise research studies conducted with them? Apart from inanimate enrichment would you provide the rats with feeding/foraging enrichment such as hay and/or produce, as well?

In my dream world I would keep my rats in relatively deep cages furnished with suitable burrowing material so that the animals can build tunnels. I would also provide them with nesting/shredding material because most rats seem to really enjoy ripping it up to make a mattress out of it, or to clog one end of a PVC tube with it.

Our sentinel rats can have whatever enrichment we think they enjoy; this includes small cardboard boxes, nesting/shredding materials and food treats. I offer them treats by hand. They get pretty excited when I visit them because they know that they will receive

treats. The sentinels are usually housed by themselves and with all the enrichment and treats/attention they receive it appears they are less stressed and calmer than other rats who don't get all this attention.

I would see to it that the rats can climb on elevated platforms or hammocks and that they have access to plastic tubes or cardboard boxes (e.g., from Kleenex tissues) to serve as shelters.

If I had a say in furnishing rat cages I would connect different cages with tunnels to make the environment more complex, and provide different substrates and different suspended enrichment gadgets along with some forage every morning in each of the interconnected cages. Some cages would have more light than others. I would mix different forage in the substrate for the rats, but I would also make sure they got regular handling so they do not get too wild!

The enrichment we give our rats varies depending on study requirements. As a basic, they all get a fun tunnel which is suspended from the lid of the cage, an aspen wood chew block, paper that they can shred, tissue paper and little plastic houses.

All our rats are caged in social settings.

We provide our rats with paper-based nest-building material, aspen wood chew-sticks and tunnels, all of which are used by the animals—although the paper strips tend to be laid one strip on top of the other to finally create a rather compact little heap rather than a nest, as mice and hamsters would do.

Our cages are manufactured from transparent, tinted material that allows the animals to observe activities outside of their home environment.

Since rats—like all rodents—are averse to open/light areas, wouldn't they prefer opaque cages to clear ones?

We did some work on rats in opaque cages versus clear cages. The rats showed no difference in most observed behaviors but some of them seemed to be more relaxed in the opaque cages.

[Blom et al. (1995) found that both albino and pigmented rats prefer cages with relatively low light intensities (<100 lx) over those with higher light intensities.]

I did a study with a colleague and compared opaque with clear cages and found that the rats spent most of the light period in the opaque cage, but an equal amount of time in the opaque and clear cages during the dark period (Cloutier & Newberry, 2010).

These observations show quite clearly that rats do prefer dark over bright living quarters. It would be quite difficult to do research with rats who are kept in the dark, i.e., their preferred illumination environment. But it is practicable to at least design their cages in such a way that the animals have free access to a place where they are sheltered from light. This can be accomplished by furnishing each cage with a covered shelter/area.

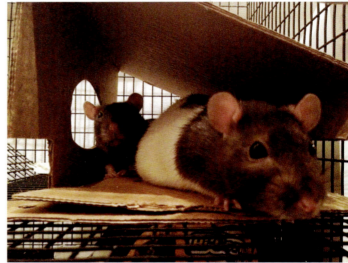

Aggressiveness in single-caged rats

We have a study of aged, single-caged male rats who have become very aggressive towards staff whenever they are handled. Aside from more frequent, friendly handling and interaction on a daily basis, does anyone have a suggestion as to other means in which these rats could possibly take out their aggression? We do not currently have enough staff to go in and handle them as much as they would need, and I believe it is rather late in the process to begin anyway, since all they want to do is attack anyone who tries to touch them.

I would try offering each rat a favored food treat before and after every handling.

In my experience, rats tend to become aggressive if they have been group-housed and then become singletons.

Rats who are alone are very unhappy and—not surprisingly—can be cranky. I have found that if a rat acts aggressively and that causes the handler to stop, the rat learns pretty quickly that the aggressive behavior does pay off.

When dealing with these types of rats in the past, I only removed my hand from their cage when they ignored it and did not react to it. Of course one has to be pretty quick not to get nailed with those big old rodent teeth, but the animals usually give you a warning before they bite. I don't normally use gloves but in this case I do. It can be a time-consuming process to gradually calm such rats down and make them feel more at ease when they need to be handled. It certainly is worth all the effort to ease their aggressive motivation which is, in my opinion, a result of the species-inappropriate solitary living conditions they are subjected to.

If the rats became aggressive suddenly, you might want to check for potentially stressful changes in their environment. I remember rats who all at once turned aggressive, and when we checked carefully for potential environmental stressors noticed a leaky water tap dripping in a metal sink. The rats became their old friendly selves once the leak was fixed and the water dripping had stopped.

Anything in the living environment making a repetitive or continuous noise has the potential of irritating rats so much that they become aggressive. Example of such noises are:
› exhaust grill or vent rattling;
› an HVAC [heating, ventilation, and air conditioning] unit with a squeaky belt;
› loud construction activity; and
› a door closer that doesn't work properly and allows the door to slam whenever a person enters or leaves the room.

We had some old rats a while ago who had to be singly housed because they were assigned to a telemetric study. These rats got very big and grumpy. Taking any free time to go down and handle them made them more friendly.

Species-appropriate housing for mice

How would you design the housing for ordinary mice, keeping in mind that the animals are not supposed to develop any so-called unwanted behaviors such as stereotypical locomotion—including jumping and flipping, bar/wire-gnawing and barbering? Please no utopia cage; its design has to be realistic and take the given constraints of a profit-focused, but at the same time science-oriented laboratory into account.

My design would involve a bit more space than the standard cages provide, especially for the larger strains like CD1 mice, and if that were not possible then only three adult mice per box instead of the traditional five.
  I would try to design the caging so that the nesting material could not come in contact with the water sipper tube and cause flooding; this would enable me to provide several different kinds of nesting material (including autoclaved hay) so that the mice could build a proper nest.

RODENTS 35

The mice would get hardwood blocks for chewing and two shelters—a paper-based one and a light-plastic one—so that there are plenty of places to hide from people and, if necessary, from cage mates.

I would love to see a doughnut- or U-shaped cage. Mice have a strong urge to stay close to walls and they are stressed in open spaces [hence the famous anxiety-inducing Open Field Test (Hall & Ballachey, 1932)]. Why do we give them cages that are mostly just open space?

I would develop a cage system with areas for the mice to truly tunnel and dig as they desire; this way they could make good use of the unspent energy some mice would otherwise put into stereotypical activities.

I designed a mouse cage that is stacked in a rat cage, with a single hole drilled into the bottom of the mouse cage. Here is why this simple cage refinement works very well for the mice:

› They can escape from threats in the upper mouse cage down to the bottom of the rat cage.
› They can hide their young in the bottom cage.
› They can more effectively thermoregulate, as it is warmer and more humid in the bottom section than in the top section of the enclosure.
› They can dig and create a real nest.
› There is lots of wall surface for the mice to be in contact with.
› The mice show no observable stereotypical behaviors in these cages.

Here is why this cage might not work for some facilities:

› It may increase bedding costs.
› Mice can be hidden from direct view.

This is a wonderful set-up. I wish all mice could be housed this way.

I think this double-decker cage is a great idea not only for mice but probably also for rats.

We have cut out half of the bottom of a large rat polycarb cage and stacked it into an intact cage. The wire lid then fits on top with food and water just like a standard cage, only it is much taller. The rats use the bottom cage to sleep and hide and enjoy climbing onto the top cage.

Bedding and nesting material for mice

I was wondering what types of nesting material are utilized by other facilities. We currently use Nestlets, but are considering other, less expensive options. We did a trial with Nesting Sheets, but there have been some concerns about the quality of the nests. I would love a product that is cost effective, but offers the mouse the option to construct a decent nest.

My facility uses mostly Nestlets, but I've found that the mice create much better nests with Enviro-dri.

It is my experience that mice prefer the Enviro-dri over the Nestlet. With the Enviro-dri the mice build a fine nest that they usually keep so tidy that you can transfer the complete nest when changing the cage, without having to add extra Enviro-dri. This material has one minor disadvantage in that it makes the cages harder to check, especially with the mice who don't like coming out of their cozy nest for the daily health check.

Our mice get iso-BLOX as standard nesting material along with some paper towels, Enviro-dri and Nesting Sheets.

We have Nestlets in every mouse cage, and it is up to the attending care staff to add Enviro-dri, facial tissues, and/or paper towels.

   My personal experience is that Nestlets combined with paper towels or Enviro-dri make the best nests. The mice seem to use the Nestlet as the foundation, and then weave the crinkly paper through.

We use regular paper towels, Nestlets and Enviro-dri (crinkly paper) and hay.

RODENTS   37

We want to give our nude mice Nestlets but wonder if the material is safe for them. Does anybody have experience in this matter?

I did a study and concluded from it that nude mice should not have Nestlets (Bazille et al., 2001). The fibrous material of which Nestlets are made disintegrates easily and gets into the animals' eyes when they groom their face. It builds up in the lower eyelid and causes conjunctivitis. [Here is the abstract of the quoted article: A colony of Hsd:Athymic Nude-nu mice was found to have an increased prevalence of conjunctivitis. It was theorized, because Athymic Nude mice lack the normal fur, i.e., guard hairs, and eye lashes, the fibers from Nestlets can easily become embedded in the conjunctiva and periorbital tissues of the eye, predisposing the eyes to chronic irritation and subsequent infection. After treatment, conjunctivitis resolved in the mice housed without Nestlets, but improved only slightly for the mice housed in boxes with Nestlets present. As a result of these findings, Athymic Nude mice are now given paper towels as cage enrichment instead of Nestlets.]

We also stopped using Nestlets with any nude/hairless mice but provide these animals with Enviro-dri without encountering any eye problems.

These results surprise me, as I have given nudes (both NCR and NU/NU) Nestlets in the past and have never noticed any problems.

I have seen both sides of the Nestlet issue. I would say 10-20% of nudes get Nestlets fibers in their eyes. The number is higher if the Nestlet is autoclaved. I have used paper towels instead, which seem to have relatively long fibers that do not compact into eyes.

Can anybody share experiences with Shepherd Shacks as shelter/nesting material for mice? Compared with Nestlets,
> *how attractive are the shacks for mice, and*
> *how practical are they for the attending care personnel?*

I've used them in the past for some sensitive breeders. The mice seemed to really enjoy them (breeding success improved vs. just a Nestlet). Typically they would chew up a portion of the shack to make a bigger nest but still keep a house-sized piece covering the nest, which they constructed with chewed-up flakes of the shack and the Nestlet. It's true, the shack did make the individual mice more difficult to see. If the bedding is not too deep and the style of shelving allows, you can normally look from the bottom of the cage and check the animals.

The Shepherd Shack we tried was a bit too tall and we had to position it very carefully to make sure that the paper-based material did not come into contact with the water bottle, causing the water to absorb into the shack and possibly cause flooding of the cage. I believe the company now makes a shack that's not so tall but has a dome-shape which makes it easier, probably, to place it in the cage without that risk.

I also used the shacks for my pet gerbils and they absolutely loved chewing them up and using the material to help fortify their nest.

We use the UK equivalent of the Shepherd Shacks and the mice love them, as they can fashion them to suit their preferences—extra exits, etc.—but they do require that you frequently have to lift the shack to check the mice, thereby disturbing them quite a bit.

We have tried Nestlets and found them totally useless, as our mice just sat on them and made no attempt to turn them into a nest.

We also supply shredded paper for nesting and have now gone over from the Shepherd Shack to a combination of plastic shelters and nesting paper for the majority of our rodents.

We have tested C57BL6, BALB/c, C3H, and DBA mice and found that all strains show a strong preference for a paper-based shelter [Shepherd Shack] over a plastic shelter [Mouse House], probably because it gives them much more opportunity for hiding, gnawing and playing—turning the structure upside down and moving it around; they even worked in order to get access to a paper-based shelter (Van Loo et al., 2005).

For those who use shredded paper as nesting material for mice, approximately how much do you give per individual so that it is not an obstruction during cage checks? We have different sizes and strains of mice, so it's a challenge to provide sufficient quantities

Did you have any problems with technicians being unable to do head counts because of the shredded paper? If so, how was this resolved?

In order to get the beneficial effects, at least 8 grams [0.3 oz] of material is needed. This is definitely enough for the animals to build a complete nest that totally shields them from sight. We have found that we do need to open the nest to do pup checks.

It is also my experience that mice make a wonderful nest with 8 grams of paper-based nesting material.

My most preferred nesting material for mice is the good old paper towel torn in half for a small mouse cage, two sheets for a large mouse cage.

The animals are counted at cage change. If they have to be counted before that, you may need to remove the cage and then look closely to be sure you can see everyone in the cage.

We have found that a lot of our mice dig down to the bottom of their cage and build the nest up from there. When you remove the cage from the rack, you can see and count all their little bodies from the bottom of the cage without disturbing the nest.

I take care of about 1,500 ventilated cages. I do not find a large nest an added burden at all. If I need to observe the animals I just pull out the cage a bit and look from underneath. If I see movement then I assume all is well. The mice get a close health check during the cage change.

When you change mouse boxes, do you transfer the nest? We change boxes twice a week, as we have static caging. I usually tell our animal care staff to transfer at least some of the nest but it is usually pretty soiled. Is it stressful for a mouse to have to rebuild a nest on a regular basis?

It depends on the animal and the condition of the nest. If there are pups, I try to leave the nest as intact as possible, and I always throw a little bit of dirty bedding into the new cage and scatter it over the floor so it doesn't smell like a completely new cage. This has cut down on male fighting. If the nest material is really scattered and soiled such as with juvie mice, I only save a very small amount for the new cage. Mice are olfactory creatures and scent means a lot to them. I think the continuity of their scent in the new cage

functions as a kind of stress buffer.

Building a new nest in such a relatively familiar environment is unlikely to stress the mice. I think the drive to build the nest is always there. If I keep adding nesting material, the mice will continue building and modifying their nest; they never seem to ever be done with it. Building a nest is probably an intrinsically satisfying activity of which mice don't easily get tired; they just start stuffing the material in the food hoppers and around the sipper tubes or in a corner. So I would say that building a new nest or rebuilding an old one is unlikely stressful for mice; it just seems to give them a satisfying job to do. Did you ever notice how excited a room gets after a cleaning, with all the mice working on their new nests? I love spending a couple of minutes when I'm done, just watching them get to work. I'll even rip the Nestlet apart if there are several mice so that they all get a little piece and can join in the nest building frenzy.

Always transfer the nest! I mean the proper nest, not the sawdust—as this might be soiled with urine. Odor cues of urine-soiled bedding and odor cues of old nesting material are not the same; they affect the mice differently. Male mice show much more aggression in a completely clean cage or in a new cage with a handful of soiled sawdust from the old cage than in a clean cage in which the old paper nest has been transferred (Van Loo et al., 2000).

I've heard that taking a small, unsoiled portion of the middle of the nest is best as that smells *like home.* I add an extra clean piece of nesting material as well, to help the mice rebuild the soiled portion.

I transfer as much of the nest as possible provided it is not too soiled.

We also transfer the whole nest or any other shelter if it's not unduly soiled.

In your experience with mice, would you say that mice have a preference for a particle size and structure of the bedding material?

We presently use two types of bedding, Beta-chip (small hardwood chips) and Alpha-dri (small square chips of alpha cellulose).
    The disadvantage of Beta-chip is that it is very dusty; the cage wire lids and micro-isolator lids must be changed more frequently than usual. I need to wear a mask when handling this substrate a lot.
    Alpha-dri is more absorbent and not as dusty, but it is harder to tell if a water bottle has a slow leak, because there is no change in color of the bedding when water is present. The cage looks dirtier than with Beta-chip because it is white, and you have to get over the urge to change it before the scheduled time.
    Can't say I see a preference for either of these two bedding materials with the mice.

Mice probably prefer bedding that they can burrow in, something that doesn't shift but will form tunnels. Generally we use softwood beddings that are kiln dried and therefore highly absorbent; it is a relatively large particle-type bedding. Wood shavings are also suitable, but they can create problems with vacuum systems of bedding removal from dirty cages.

I wrote my dissertation about bedding preferences of group-housed female mice;

I only tested wood product bedding, not nesting material. *My* mice clearly preferred wood shavings over wood chip bedding. [Dwelling times on the particular bedding structures were statistically analysed as a parameter for bedding preferences. In all three test combinations, a highly significant shaving preference was detected. On average, mice spent 70% of their dwelling time on the shavings. This preference was more explicit during the light period and in C57BL/6J mice. The relative ranking of the bedding structures was: shavings>>coarse-grained chips>medium chips=fine chips. By means of these results, a shaving structure as bedding can be recommended for laboratory mice, whereas fine chip structures should be avoided (Kirchner et al., 2012).]

Suppose you are a mouse who is genetically not altered and you could chose the nesting/shelter material and the bedding material for your traditional mouse cage; what kind of nesting/shelter material and what kind of bedding would be your preference?

My preference would be a Shepherd Shack along with a whole Nestlet and a bit of shredded paper. I could make a fine nest with this material but also hope that I can build it away from the water sipper tube so as to avoid flooding of my cage.

I would choose a Nestlet and shredded paper, as I would be able to cater to my natural drive to build a nice nest.

An igloo with the satellite running-dish attachment would be on my wish list. I see myself as a running enthusiast even as a mouse … LOL. I would also like a Nestlet with some crinkled paper for nest building and Bed-o'Cobs for bedding. I think this all would make for a very cozy home.

I would like to have wood shavings at least 10 cm [4 inches] deep for burrowing, and enough tissue paper plus some wood wool for nest building.

This falls outside your requirement for a traditional cage, but if money was no object and you did not want to catch me easily, I would like about 12 inches [30.5 cm] of peat in which I could build tunnels and chambers like I do in the wild!

As a side note, inbred mice who have not encountered a digging substrate for at least 20 generations will take just about 30 minutes to build a perfect burrow when they get access to a generous amount of peat. Mice are highly motivated to burrow in suitable substrate; burrowing seems to be a behavioral need for them (Sherwin et al., 2004).

This reminds me of a little mice colony that had dug numerous tunnels through the fiberglass insulation in the floor of our old cabin (shown below).

Now that we know what kind of bedding and nesting materials would make a mouse happy, the question arises: what type of bedding and nesting material would you recommend for mice living in individually ventilated cages [IVCs]?

A Shepherd Shack and a half Nestlet with a bit of wood shavings would be practical and safe, and it would allow the mice to express their species-typical need to build a snug nest.

I would recommend tissue paper and a Shepherd Shack; these items allow the mice to build a suitable nest, and the shack protects them against the drafty environment of 50 or more air changes per hour.

My preference would be a 0.25 inch [0.6 cm] corncob bedding along with paper towels for nest building and a tube, hut or house serving as shelter.

Usefulness of the term *enrichment*

It seems to be accepted language to use the term environmental enrichment *for species-adequate nesting material. Is it really fair to speak of enrichment when the so-called enriching material is a biological necessity—rather than a generous luxury—for the subject's well-being? After all, a mouse has to build a nest in order to be protected.*

I couldn't agree more. We do need better terminology to separate behavioral requirements/needs from environmental enrichment/fun. Sometimes I can't get my head around that term *enrichment* and think

it only applies to stuff given to an animal that is not really a necessity but a kind of entertaining toy. A good example is offering a rabbit a stainless steel bowl to throw around. It's probably not necessary for the well-being of rabbits but they sure enjoy making noise when it is their idea.

I also don't like the term *enrichment* which, indeed, is suggesting luxury. *Environmental refinement* is perhaps a more appropriate term.

We should do away with the term *environmental enrichment* and replace it with *essential enhancement*.

I think the history of how we cage animals for research has made the barren cage the standard, validated by studies that are repeatable and have been documented over the past 50 years. To use the claim that nesting material is a necessity does not hold water with the older generation of administrators and scientists. The good news is, those folks are retiring and the younger generation is already used to working with animals who live in enriched living quarters. For them, nesting material for mice is a standard supply for every mouse cage.

So, things ARE changing; it's just a matter of time!

Preferred nesting location of mice

Mice tend to prefer nesting in the rear, rather than in the center or front section of traditional cages. What could be the reason for this preference?

It's darker in the rear.

Probably because the rear of the cage is relatively dark, hence secluded.

Mice are nocturnal animals, so it is a natural response to nest in the darkest area of the cage, which is normally the rear section of it.

Logically the nest is away from light and traffic.

Agreed, away from light and traffic, and also away from the water bottle and the food hopper.

How is the situation in individually ventilated cages?
Will mice, who prefer to build their nests away from light, satisfy their preference in IVCs, or will they adjust their nest location in relation to the location of the enforced air supply in the cage?

I care for two vent racks with various strains of mice and have noticed that most mice build their nest away from the air flow vent.

This is also my observation; the mice avoid the air stream by choosing another resting place in the cage, or by piling up walls of sawdust around the nesting place or use provided nesting material or a movable shelter as a windshield.

In line with this is the observation by Scales & McDonald (2011) that "61% of mice housed in static caging preferred to nest in the rear of the cage, compared with 49% of mice in low ventilation [30 ACH] caging, and only 14% in moderate ventilation [70 ACH] caging." Obviously, mice don't like to be constantly exposed to a strong air stream, so they build their nests away from it, even if it implies away from the relative dark and undisturbed rear section of their cage.

We don't have to set the air changes so high that the mice are living in a wind tunnel.

Mouse cage changing

When changing mouse cages that contain breeding mice with offspring, I first move one parent to the clean cage, then the litter and only then the rest of the adults. The pups seem calmer when I do this, especially at the popcorn age. I like to think that the familiar smell of the parent in the new cage is the reason for this. Does anyone else find this?

If you transfer singles it takes more time and the mice scurry all over the place; that's not good. We just try and scoop 'em all up at the same time. If a little soiled bedding goes along that's a good thing. The mice settle right in.

I agree; when you scoop up several mice at the same time and transfer them into a new cage, the animals are calmer than when they are transferred one by one. Also, if you scoop up a few adult mice, they are very unlikely to bite you whereas a single mouse will experience a lot of fear and, therefore, be more defensive and ready to bite.

The multiple-mouse idea is great but it gives me less control over the mice. I had the experience several times that some of them jumped on to the transfer station, then panicked and jumped down to the floor during the process. When this happens the whole group of mice has to be sacrificed. Researchers tend to get a little bent out of shape about this, especially if the mice are important or expensive. It bothers me because it really hurts the mice.

Treating UD [ulcerative dermatitis] in mice

We are having lots of UD cases at our facility and find it difficult to get cases resolved. Can anybody share experiences on how to treat UD with reasonable success?

We have had pretty good success using a chlorhexidine (1:8) solution three times a week along with trimming the nails.

UD is quite commonly seen in C57Bl mice, especially GM mice with that background. Baytril helps but when you stop applying it, the dermatitis starts again. UD is very difficult to treat successfully.

I have noticed over the years that a lot—not all—of UD cases start with barbering on the neck and back and then progress from there. It is most common in C57s.

Even single-housed C57Bl mice develop UD. I would treat the affected mice with Baytril, at least until the end of the study they are assigned to.

A lot of our recent UD cases, referred to me for a behavior consult, aren't related to barbering. The typical barbering patterns (facial baldness, whiskers missing, and bald patches) are absent and the mice have been observed scratching the UD areas with their hind feet. So maybe it isn't barbering (using incisors to pull hair from self or cohort) but self-scratching with dirty sharp nails that leads to UD in association with a genetic predisposition.

I have to say that in my somewhat limited experience with this, none of the UD cases were the result of barbering, as almost all of these mice were single-caged. Those that lived in groups were separated, but the condition did not resolve. The cases I have seen look more like compulsive scratching, first of the ears and neck area, and then of the hind-end flank area.

If ulcerative dermatitis is mainly a result of a behavioral problem, then trimming nails and adding mice-adequate enrichment may be the most appropriate first treatment attempt.

Foraging enrichment for rodents

If you are in charge of rodents, do the animals receive any kind of foraging enrichment such as small food items mixed with fresh bedding?

Ours do not. It is on *the list* of things to look into; I wish we could make it possible for our rodents to engage in foraging activities.

We don't do that either, but I am going to try getting our rodents some certified treats that we can mix into the clean bedding. Presently I throw a handful of their standard food pellets onto the cage bedding rather than into the feeder. The rats especially seem to like this; they get hold of pellets and run around with them before actually eating them.

At every cage change I scatter a small amount of irradiated sunflower seeds, rabbit food, or certified treats on the bedding of our rodents. I also distribute part of the daily standard food ration on the bedding; I have done this for years in many studies including GLP [good laboratory practice] studies. The amount of the extra treats is so small that it does not affect calorie intake and body mass composition. However, I always make sure that the treats for foraging are mentioned in the study protocol and signed off by the study director.

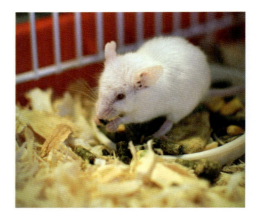

Scattering seeds or other small food items on the bedding works great for mice; they love it. The problem is that these treats may interfere with nutritional studies or experiments where body weight is an important parameter. We became aware of this some years ago when everybody was happy—especially the mice—until the researcher told me that he was surprised that the body weight of the animals was yo-yoing so much. We found out that the animal caretaker had given the grains always at 4 p.m. before he went home. In just a few days the mice noticed this predictable routine. They liked the seeds more than their standard food, so they waited until 4 p.m. and then started to forage and eat the grain, and continued eating pellets after they had finished all the grain. We had to prevent this—for the sake of the study—and asked the animal staff to give the mice the grain in small portions at different time points throughout the day. Animals always surprise you!

Shelters for rodents

If you are working with guinea pigs, do you think that the animals benefit when they have access to a hiding place? If so, what kind of shelter(s) would you recommend?

My own guinea pig at home enjoys her chewable shelter—it looks like a hollowed-out log—and she will also go under hay. When I first got her, she used the shelter a lot and slept in it every night. Now she is so relaxed that she hides in it only rarely, but she runs through it at times. I would think that guinea pigs in the lab are relatively apprehensive of people so they would benefit from a hiding place, not a transparent one that care personnel would prefer so that they can monitor the animals, but an opaque dark one in which they can really feel secluded and safe.

All our guinea pigs have access to little plastic huts. For group-housed animals, we provide enough huts so that they all can find a safe place in one of them. They definitely retreat into the shelter when they are spooked, especially the newer/younger ones who are more skittish; I do think guinea pigs truly benefit from having a secluded place where they can take refuge. I like that the huts are open on both ends so that the animals can escape from each end if needed.

[Walters et al. (2012) found that pair-housed male guinea pigs "with a hut had significantly lower fecal cortisol concentration than pair-housed animals without a hut." This indicates that a refuge can serve as a stress buffer for guinea pigs.]

I think a hiding place for guinea pigs is a *must*. These are typically nervous animals and whether they are single- or group-housed, they do require access to a shelter.

It is my experience with group-housed guinea pigs that the animals like hay more than any other substrate. Is it practicable on a regular basis to provide single-caged or pair-caged guinea pigs with an amount of hay that is sufficient for them to hide in/under?

I believe the benefit to the animals outweighs all potential practical issues, such as extra time investment for cage cleaning and hay distribution. Hay is not costly, and it takes just a few minutes a day to give each animal a handful. When changing the bin, I have to dispose of the dirty shavings anyway, so some hay along with that doesn't take up any more time.

Hay is indeed the best so-called enrichment you can offer a guinea pig; it can be used for hiding, playing, nibbling and foraging. Unfortunately, some institutes will not allow hay behind the barrier without autoclaving, which makes it brittle and more sharp, hence potentially dangerous for the animals' eyes.

We also consider hay as an essential element of guinea pig housing and care. Our guinea pigs get a large slice from the hay bale on a daily basis; this is sufficient for them both to forage in and to hide under. To distribute the hay takes only a couple of minutes. We used to autoclave all our hay but now purchase it irradiated, but only because it's easier to store. We have not experienced any problems with eyes irritated by hay, but straw can be a problem unless wheat straw is used—barley straw has awns that can penetrate soft skin and eyes.

If a restricted-nutrient study does not allow for the provision of hay, we give our animals wheat straw instead.

All our guinea pigs also have access to shelters, but they clearly prefer the hay or straw as hiding places.

[Hay not only is a great hiding place, but it has additional side effects that are quite beneficial for guinea pigs who live in research labs. Gerold et al. (1997) found that providing guinea pigs hay—an important source of crude fiber—substantially reduces hair loss resulting from hair-pulling-and-eating. Cozen (2006) noticed that groups of male guinea pigs were less difficult to handle and were less aggressive among each other when they received hay than when they had no access to hay.]

Tubes—cardboard or plastic—are often used as shelters for rodents. Typically such tubes are open at BOTH ends. The biologically natural burrow of rodents is closed at one end, probably providing the animals a relatively greater sense of security.

I am wondering, would rodents in the research lab prefer tubes that are closed at one end over tubes that are open at both ends?

I've noticed that rats tend to plug one end with their rear so that any attack does minimal harm. I can say that they do prefer one-end-open tubes over both-ends-open tubes when they have the choice.

For rats, the closed tube may be more comforting. In our cages the open end is usually positioned against a wall of the cage, so the two exits are not a benefit anyway. It is my experience with rats that this circumstance does not create any special aggression problems, so a tube with only one exit is probably just fine for them.

The situation is very different with mice. The more exits a shelter has the better it is for them, especially at cage change when they get aggressively excited. A shelter without escape options is likely to lead to serious beatings. Dominant mice can be merciless when the victim does not get out of their sight.

Mice would probably use a tube with only one entrance as an ambush point; therefore, chances are that such a shelter would increase aggression within the group.

We have performed numerous preference tests with different types of shelters. What I recall

very clearly is that mice, given a shelter with one opening, very often would not sleep inside the shelter, but just outside. I have always presumed that this was because the mouse did not have a way out if trapped. In nature, I believe mouse burrows do have several flight routes. Another observation we made when testing Shepherd Shacks (cardboard nest boxes with one opening) is that mice quickly transform the shacks into a cardboard nest box with two or even three openings.

Wheel-running in rodents

Wheel-running seems to be a very attractive activity for caged rodents. Since stereotypical behaviors develop during early life, I am wondering if rodents exposed to running wheels since birth show less stereotypical activities when they are adult than rodents raised and kept in barren cages.

To my knowledge this question has not been addressed in any published article.

Could it not be that wheel-running itself can develop into stereotypic behavior? I've always found it difficult to decide whether mice engage in wheel-running (a) freely, because they enjoy it, (b) compulsively, as a repetitive stereotypic behavior, or (c) as a way to react/escape after a stressful event such as being handled by a person. Maybe all three phenomena occur from time to time!

What I have seen in BALB/C mice is that when being exposed to a wheel for the first time, they will run but also play in the wheel for several days and after that will stop playing but keep running on and on. That might tell us that they do in fact develop the wheel-running into a stereotypy over time.

Do we really need to worry about stereotypical locomotion, such as running in a wheel, running in circles, pacing back and forth, somersaulting, bouncing and back-flipping (a) when these activities are performed by confined animals who don't have enough room to express their species-typical drive for exercising, and (b) when these activities are non-injurious?

Of course we should worry, as stereotypical locomotion such as wheel-running is an abnormal behavior. To me it is questionable whether running in a wheel in a small cage for hours can be beneficial, as the animals actually lose weight. In their natural environment, rodents are probably very cautious with their energy expenditure, in order to be fit for important things such as foraging, exploring/patrolling the boundaries of their territories and breeding. For that reason I consider wheel-running done for hours without resting or eating as abnormal behavior.

If caged rodents run voluntarily in their wheels to the point of exhaustion, doesn't that show us that the artificial environment in which we confine them is not right? Seems to me that excessive running in a wheel is a

desperate *normal* attempt to somehow cope with extremely *abnormal* living quarters. The rodents implicitly tell us: "Look, the cages you confine us in are not suited for us, please use your brains and refine them so that we can behave in a more species-appropriate way!"

I think animals in research labs try to make the most of the environment they are restricted to.

 I probably sound really silly, but I am a busy person like many of you. I commute four hours a day, work full time, have a family to care for (two small children and a husband), a home to maintain, and no time left to do the one behavior I long to do every day: go to the GYM. So what do I do? Same thing as the rodent on the wheel: I made the most of my restricted time and environment and bought a treadmill. So, I run like a maniac in whatever small amount of spare time I can come up with until I too am exhausted.

You very succinctly made it clear what our present discussion here is all about: do the best you can to get some exercise, even if it implies running in a wheel [or in a treadmill].

I still don't think that running in a wheel is comparable with running out there in the wild. Rodents in the wild walk and run over quite a distance to find food or sex partners; they climb and play but they don't spend energy in biologically useless running. So from that viewpoint, a cage with possibilities to climb, hide and play might meet their needs for exercise more than providing them with a running wheel.

I have problems with the implicit assumption that running in a wheel is more abnormal than running freely. In both situations, the rodents express their biologically programmed drive to make use of their little legs (a) either in the unstructured, species-inadequate environment of the cage or (b) in a naturally structured, species-adequate environment where they are free.

Are mice, rats or hamsters in the artificial environments that we provide them in the research lab actually *normal* animals? I have heard that rodents will travel several miles at night in their natural habitats. How can they do this in unstructured cages where they have no choice but to travel round and round, over and over again along the restricted perimeter of the small cage? I would say that this is truly

abnormal stereotypical behavior. When we give these animals the opportunity to travel, so to speak, over long distances in running wheels, we may actually turn them into more normal rodents, hence better research models.

I am wondering: will the animals make use of a running wheel in the characteristic stereotypical fashion when we keep them in spacious, well-structured cages that allow them to engage in foraging activities, burrowing, building a nest, seeking shelter, climbing, checking the environment from a lookout, and running from one functional location to the other?

We have group-housed mice who have a spacious cage with a variety of enrichment items including a Nylabone, an igloo with running wheel, a plain igloo, a tunnel, and nesting material. There is pretty much always at least one mouse on the running wheel. It is constantly in use, sometimes up to five mice running together in it. It is fun to watch them run, and it does not appear to me that this running in the wheel is an abnormal behavior. All of the other enrichment items are also used, but even with the variety provided along with the spacious living quarters, the mice still use the wheel heavily.

Your mice *want* to run in the wheel; there is no real need for them to do so as they are not confined in a too small cage with nothing to do but run compulsively in a wheel. They have enough room to run around in a normal fashion and they have access to enrichment items that they also make use of. If we would classify the wheel-running of these mice as an abnormal behavior we would also have to classify the gnawing of the Nylabone, the climbing on the igloo, the running through the tunnel, the hiding in the tunnel and the building of a nest as abnormal.

I agree, wheel-running is more an adaptive rather than an abnormal behavior. The mice make use of their environment to suit their needs. I've had pet rats and hedgehogs use running wheels even though they were kept in large, well-structured interactive cages.

Obviously, it would be too simplistic to categorically label wheel-running in rodents as an abnormal behavior without first taking the quality of the animals' living quarters into account.

I firmly believe there are times when mice do indeed *want* to run in the wheel or sometimes simply go for a ride. We had a study going on that provided group-housed mice with very large cages containing a wheel. The animals were handled about once a week and otherwise left alone so that it could be seen if they would run by choice. Some of the mice loved to run in the wheel. We would see them run on and off during the day and the research group reported that night activity was increased.

    My personal favorite moment was when I peered into the room and discovered one mouse *hitching* a ride in the wheel. He sat on the bottom and allowed one of his cage mates to do all the work! He would ride up and then slide down only to ride up again and then slide down again. First, I thought it was simply one of those who got caught sleeping in the wheel until I saw him doing it again and again a couple of days later. I swear if mice could smile, he would have had an ear to ear grin plastered across his little face!

Rodent enrichment – census

I'd like to take an unofficial poll on LAREF to see how many of your institutions are using some form of enrichment in the rodent cages and, if so, what percentage of all cages at your institution are enriched?

At our facility, 100% of all rodent cages are provisioned with some form of enrichment.

All our rodent cages have enrichment unless there are exceptional research-related reasons for keeping animals in barren cages. Our inspectors expect that all our rodent cages are enriched, and they will write it into their reports whether or not enrichment is in place.

My institution provides at least one enrichment item in every rodent cage.

At our facility we add a tunnel, two paper strips (1 x 10 in / 2.5 x 25 cm) and a Nylabone to all of our rat and mouse cages.

All of our rodents get gnawing blocks. In addition we give our mice cardboard tubes and nesting material; our rats get Enviro-dri.

All our rodents have access to nesting material (shredded paper, tissues, or Nestlets) or to a pre-made shelter (Shepherd Shacks or igloos). In addition to this, wooden gnawing sticks are provided for the rats and the gerbils.

At our institution, also, 100% of all rodent cages contain some form of enrichment.

The same is true for our lab.

All our rodent cages have nesting material; at cage change we add a food treat, a paper-based shelter or a chew stick. The only exception to nesting material is behavioral cages on monitors where the computer would lose the target animal in the nest. In that case we do have enough contact bedding so that the animals can make at least a little dent in it to serve as a nest substitute; they also get sunflower seeds scattered on their bedding every day so that they can get busy with foraging.

There is no good reason to offer caged rodents no enrichment, even in pivotal-phase GLP studies.

We provide paper-based nesting materials, wooden chew-sticks and plastic shelters for all of our rats and mice unless there is an IACUC-approved scientific justification for one or more of these items not to be used.

All our rodent cages—except breeding cages with pups—are furnished with a species-appropriate shelter. In addition, all mice get nesting material and all rats get a wooden chew-block.

I hope someday these items will be classed as *behavioral necessities*—rather than enrichments—and be the standard furniture of all rodent cages.

I agree, rodents are probably not regarding nesting material and a shelter as *enriching* their cages but as necessities for engaging in activities that have significant survival value for them.

Training rodents to cooperate during procedures

Can you train rodents to cooperate with you during certain procedures?

In one of our smaller facilities that usually only has a few cages of rats we were able to train Long-Evans and Sprague-Dawley rats to self-cage change: the tech puts the clean cage adjacent to the dirty cage; then, when the rats approach the side with the clean cage, the tech gently lifts the rats' rears so they can hop over. The animals quickly learn to jump over on their own; our smaller rats get a lift for a bit longer but only because they aren't long enough to make it up and over. We don't reward the rats with treats until the cage change is complete and all are over. The cages are changed 2–3 times a week (static shoebox style cages); usually within two weeks at the longest, the rats are changing themselves. The time required for self-cage change and manual cage change is about the same.

I did something similar during grad school when I had a colony of really old, crusty, cranky rats who had been frequently handled when they were young. The cages needed to be changed 2–3 times a week, and I had gotten sick of being bitten when picking the rats up manually in order to transfer them into the clean cage. It turned out to be quite easy to get them to cooperate during cage change: move the clean cage next to the dirty cage and then gently help the rats to get over, and reward them after everyone has made it into the clean cage. I did have to place a little box close to the adjacent sides of the two cages to assist some of the rats who couldn't climb up in order to hop over. They all learned to move from the dirty to the clean cage. Some days, cage change took a while; it didn't really matter, but it made life less stressful not only for the rats but also for me.

One is tempted to wonder why this simple training exercise of rats is not a standard procedure.

When working with rats it quickly became very clear for me that they are very attentive and smart animals who can be trained—using basic and gentle skills along with a food reward—to cooperate with me. For example, rats on a gavage study learned to literally open their mouths and allow me to insert the feeding tube without struggling. I remember an immunosuppression study in which rats were receiving a subcutaneous injection once a day over a period of one year. Since quite a number of animals developed lesions at the site of injection as a result of the immunosuppression, the principal investigator told me that he usually adds 12 extra replacement rats for his study. I suggested that we would give the rats a food reward after each injection so that they could learn to associate the daily injections with a pleasant experience, hence accept them without fear. We would only need one extra potential replacement rat, just in case. The PI was very excited to hear this because it meant treating 11 less rats for 365 days, hence

saving money and resources. He did ask me if I was sure that it would work. Well, it worked just fine. I chose dog food as the reward for each injection. The rats loved it and readily learned to cooperate with the injections. We had zero lesions and all animals in the study survived the year! So yes, rats can be trained to overcome injection-associated stress.

I reward mice and hamsters with sunflower seeds or rabbit chow after each handling procedure but have yet to notice any positive change in their stress reactions to treatments.

[Rats are smart animals who have also been trained with success to cooperate during blood collection (Shyu et al., 1987), saliva collection (Guhad & Hau, 1996), and oral dosing (Huang-Brown & Guhad, 2002; Rourke & Pemberton, 2007).]

A few years ago, an animal technician gave me a tour of an institution in New Zealand. When she opened the door to a mouse room, all of the mice came to the front of the cages immediately! She told me that she had to give these mice daily intraperitoneal injections and started to give them a reward after the treatment. The mice were easy to handle and, obviously, came to the front of their cages in anticipation of the reward after being injected. I could hardly believe what I saw and asked which treat she used and she said small chicken pellets! Upon returning to my institution in the Netherlands I applied the same trick during a study with mice and yes, it worked; the mice loved the chicken pellets and, in return, allowed us to give them daily injections!

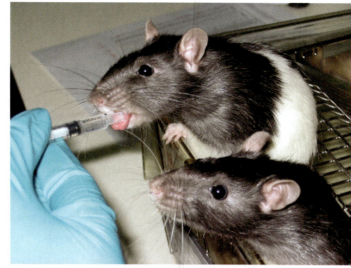

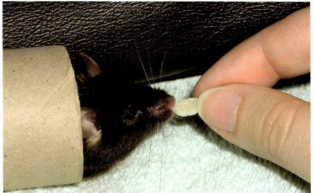

RABBITS

Environmental enrichment for rabbits

What are the options for providing single-caged rabbits commercial or custom-made environmental enrichment gadgets that do NOT contain food, yet require no rotation because the animals don't lose interest in them over time?

Does hay count as food?

Hay counts as food. As long as a gadget contains hay, it is very unlikely to become boring for a healthy rabbit, but what about gadgets without hay or any other food item?

Chains, rattles and Kongs elicit temporary interest in some rabbits. Cardboard is more attractive and fun for most rabbits. I cut cardboard boxes into halves and give those to the rabbits. They will sit mainly on them rather than in them, and they will chew and rip them apart in the course of a few days. At cage change I simply throw the boxes out and exchange them.

I used to give paper bags to my rabbits a few years ago every Sunday afternoon. The rabbits would happy-hop in great anticipation as soon as the paper bag cart came in.

Cardboard boxes for sure are very entertaining for rabbits!

I spend a lot of time removing the plastic backing from pan liners that we use at our facility. The liners are of tissue paper-like quality, and the rabbits make very elaborate nests out of them.

Small soda bottles partly filled with pelleted bedding and the lids glued on are great hits; the rabbits don't seem to get tired of pushing these rattling cylinders around their cages.

A hide box should be basic furniture in every rabbit cage; it is not really an enrichment but a necessity.

Cardboard boxes for chewing and small stainless steel bowls for making noise are never ignored by single-caged rabbits.

We use the stainless steel rings from canning jar lids. The small ones of course so the rabbits can't pull them over their heads. They don't seem to get tired of throwing and nudging these rings around, thereby making a lot of noise. There is no need to rotate the rings, as the rabbits don't lose interest in them.

That sounds like a simple and inexpensive idea. How do you clean/disinfect the rings? We move them with the rabbit and then just throw them away when they start rusting.

Our rabbits love the Rabbit Race Car. It consists of a stainless steel bolt with loose stainless steel washers and loops. I would have never thought it would be the most used rabbit toy in our inventory, but the animals get a kick out of picking them up and tossing them around, thereby creating quite a bit of rattling noise. They toss them almost immediately after they are placed in their cages and continue interacting with them for several weeks. What is interesting is that the rabbits don't show any startle response when they are tossing them around in their cages, and other rabbits in the room seem to remain undisturbed, not at all startled when their neighbors are making such a racket.

These gadgets are, apparently, a great idea. I guess they are particularly entertaining for the rabbits because of the noise they can produce. Anything will do if a caged rabbit can push or throw the gadget around, thereby creating noise; the more noise the better.

Suppose you have worked with rabbits for a long time, know them very well and are in a position to interpret their behaviors and behavioral reactions correctly. Now, when you watch your rabbits making a heck of a noise with gadgets, do you have the impression that they are in a playing mood or are they making the noise for a specific purpose? Perhaps they are simply playing and have fun moving those noisy gadgets around?

Rabbits love to make noise when it is their idea. I am sure, in the wild it would serve no purpose and would attract a predator pretty quickly, but give a rabbit in a bank a small stainless steel bowl and she or he will have hours of enjoyment picking it up and throwing it around, making tons of noise!

There's something almost comical about how much rabbits enjoy making noise. It sure seems to make them happy.

Perhaps that's the purpose behind the noise making!

I can only confirm with my experiences that the rabbits definitely have fun making noise. We've used rings from canning jar lids, bird toys consisting of hanging bells, and a collection of metal objects clipped to the cage front; all of these noise-producing items are greatly welcomed by our rabbits who are using them with remarkable consistency.

By the way, for those who house many rabbits in cages, be aware, the rooms can get pretty noisy at times when all animals are playing with these types of toys.

For me personally, I've just always assumed the rabbits were having fun and had no other motive other than to have fun. I could be completely wrong, but their frequent interaction with noise-producing toys and their body postures and attitudes while playing with these gadgets make me believe that these rabbits are happy; so even if I have to wear hearing protection, let's give them the noisy toys!

Rabbits give the impression that they are simply having fun when they play with noise-producing toys.

We can conclude from our rabbit-making-a-lot-of-noise discussion that noisy gadgets do, indeed, enrich the environment of caged rabbits by distracting them from chronic boredom. We may also conclude that the noise-producing gadgets are not necessarily also enriching for the attending care staff, but they may put up with it for the sake of the bunnies.

The published literature on the usefulness of mirrors as enrichment gadgets for rabbits is equivocal. Does anybody on the forum work with rabbits who receive continuous access to mirrors? Based on your own observations, would you recommend mirrors for environmental enrichment?

We currently give little mirrors to rabbits with behavioral issues. I have observed that the animals are using their mirror only as a rattle; they don't really look into it to watch the reflection of themselves or other objects.

Based on my experience, mirrors make good rattle devices for rabbits but aren't really utilized by the animals to see reflections in them.

I have seen rabbits using mirrors suspended on chains as noise-producing gadgets. They may also chew on them or push them around but they don't use them as primates would to check the mirror's reflections.

I am trying to put together a playpen for our rabbits. Can anyone share ideas of suitable pens and how best to furnish them?

The small, plastic step stools work very well as a platform and at the same time as a shelter. If you line several of them up, you can create a tunnel through which the rabbits will run. Bedding bags are also popular as short-lived hiding places that will be ripped apart with much noise and finally shredded into small paper fragments.

Paper floor liners are favorite toys for rabbits; they rip and tear them and turn them into burrows and, finally, a heap of shredded paper. If there's not enough room for an entire floor liner then a paper towel is a great substitute. Both items have the bonus of promoting the natural behaviors of nest making and rooting.

We use commercial exercise play pens with lids for our rabbits. When we have a mom with kits, we attach rubber mats or sheets of thick plastic to the walls with zip ties about 12 inches [30.5 cm] high to prevent the little bunnies from sticking their heads through the wires. We toss aspen shavings on the bottom. Treats are distributed in feeders or on the ground to promote foraging activities.

Commercial shelters for small dogs serve as places to hide inside and to hop on top. We prefer the playpens without a bottom, as they're easier to fold up to be then sent to cage wash.

If you have the space, you could turn a small room or a fenced-off portion of it into a floor playpen for your rabbits. Wood shavings and heaps of shredded paper with some treats added will keep your rabbits busy playing and foraging. Empty paper bags are used as short-lived hiding places that are turned into shreds. When you put some hay into the bags and then roll them up tightly, the rabbits will get a special foraging task. You can use soap barrels and either cut them in half or lengthwise for hutches and resting platforms. Multi-level contraptions are easily created by taping sturdy cardboard boxes of different sizes together. I use recycled cardboard boxes without staples and select those with minimal ink printing on them. When you cut holes in the boxes, the rabbits will use them as bolt-holes and as lookouts.

Based on your own experience with group-housed rabbits, would you say that elevated boards/shelves are (a) useful and (b) safe enrichment structures for the animals?

I would say that elevated boards/shelves are useful and safe, provided they are properly secured and are installed at a reasonable height. All our group-housed rabbits have access to elevated lookouts, and they are using them all the time. Over the years I have never had to deal with an animal who had an injury or fracture that was related to a resting board, a platform or a perch.

Rabbits like shelves and elevated places as they appeal to their natural habit of sitting on their hind paws and looking out. I have never encountered a case of a rabbit breaking a leg when jumping down from a platform.

Our bunnies have shelving and they love to lounge up there. I think it makes them feel more safe. Elevated lookouts are also used as a means of vigorous exercise, and the rabbits often jump up and down from shelves as they zoom around. So far no injuries. It would be interesting to find out if this kind of exercise strengthens their muscles and helps to maintain a good bone density that may be protective against fractures.

I was really nervous when I first put the shelves in the pens as they seemed pretty high to me—above the Lixit—but the rabbits figured it out with ease in no time. Rabbits seem to have a strong urge to overlook their environment from higher ground; this may give them a sense of visual control and safety.

Our group-housed rabbits also like to perch themselves at a higher level. We cut old barrels into halves to serve as shelters and lookouts. Occasionally we have a rabbit who will leap from the top of a barrel over the enclosure walls, but we have never had any injury resulting from this.

Housing female rabbits in pairs

Can anybody share experiences on safe pair-formation of previously single-caged female rabbits?

A few years ago my charges were single-caged female rabbits on a 5-year study. When I started working with these animals they had been there already for about 2½ years. They lived in banks and were developing hock sores and other foot issues.

Another tech and I developed a floor-housing system. We were in our own building, so we had plenty of space and were able to transfer all 50 rabbits to the floors of three rooms. The new housing consisted of collapsible dog pens with connected pet-shade on the tops for escape artists.

I arranged the does based on personalities and placed two of them as neighbors when I had the feeling that they would like each other. They could contact each other through the widely spaced bars of the pens. When they would lie next to each other consistently, were often grooming each other peacefully, and none of them showed any signs of depression, I paired them by removing the divider.

We were able to match up all 50 does into 25 pairs. There was only one pair formation attempt that resulted in a serious injury.

After my experience with those girls, I firmly believe that single-caging of does should be abandoned and replaced by pair-housing in floor-pens.

Your belief is supported by two studies showing (1) that does have a strong preference to spend time with another doe rather than alone in another cage (Brooks et al., 1993), and (2) that does are willing to push through weighted doors in order to gain access to another female companion (Chu et al., 2002).

We have had significant success introducing long-term singly housed does with each other for subsequent pair-housing. We observe potential companions repeatedly for short socialization sessions in a playpen that is new to both of them and hence unlikely to prompt territorial activities. If the two rabbits show affiliative behavior during most of the socialization periods without engaging in aggressive behaviors, we have been able to successfully pair-house them in more than 80% of cases.

[A well designed pair formation protocol for female rabbits has been described in Tech Talk (2009) by Fuller: "Potential partners were first

given an extensive 'getting to know' you period through several introduction sessions. Two females were chosen based on size and general temperament (same-tempered rabbits who are the same size are far more likely to get along) and were placed into an exercise pen together for at least 20 minutes a day for 2 weeks or longer. The floor under the pen was covered with textured cardboard to provide traction, and a disposable cardboard shelter was provided. Enrichment devices were also placed in the pen, including Jingle Balls and timothy hay cubes.

"During these introduction sessions, the rabbits were observed at all times. Expected behaviors included hyperpnoea and chasing, as well as occasional vocalization and stomping. Mounting and hair pulling were also commonly seen; females engage in these behaviors in order to establish dominance. After several sessions, if the rabbits appeared to be more interested in confrontation than exploring their surroundings, the pair was separated. If any evidence of injury or extreme stress (dyspnoea, pale ears) was seen, the rabbits were separated immediately.

"After each introduction session, the rabbits were returned to their regular housing. We swapped their feed hoppers and enrichment devices (shelters, toys) so that the rabbits would become accustomed to having the smell of their pair mates in their home cages. Rabbits were then given a timothy hay cube so that each session ended with positive reinforcement. At the conclusion of the two-week introduction period, it was obvious which pairs were able to be successfully housed together: the expected behaviors listed above became less frequent, and positive signs such as nose touching and mutual grooming were observed.

"Pairing was performed at the start of a work day, on a rack-change day, so that neither doe had the chance to mark their territory. … The divider between the two cages was pulled out halfway for the first 3 hours, which helped cut down on the amount of chasing the rabbits can do and allowed the

rabbits to spend time apart until they became used to the new experience of having a cage mate. Each rabbit pair was also given a rabbit shelter. Sheltering is especially important on the first day of pair housing, as almost every rabbit shows apprehension during such a drastic change in their housing conditions.
… If there were no unexpected adverse behaviors after 3 hours, the divider was removed.

"The pair-housed rabbits were observed several times a day for the first three days."

The author successfully formed seven compatible pairs of adult female rabbits. In almost all instances, the pairings resulted in an obvious "bond" between the new cage companions.]

Bunny nest

We do a lot of repro-tox work with rabbits who are all kept in stainless steel cages. One of our techs and I are bothered by the fact that these rabbits don't have a suitable place to build their nest; they are agitated and seem to be frustrated to not find a spot that could serve them as nest. So, we would like to add a solid floor space (like a low

tray or a floor insert) to give them a place where they can build their nest with suitable nesting material. I imagine that this would decrease their distress toward the end of the study when they approach parturition; we desperately need data to get our refinement idea implemented by the study directors.

We used the [very expensive] nesting box from Otto Environmental—a 10 x 10 x 20 inch [25 x 25 x 50 cm] stainless steel construction with removable polycarbonate floor—to breed some of the first transgenic rabbits. As long as this cage addition was specified in the study outline, our PIs had no problem with it.

Pregnant rabbits exhibit nesting behaviors as they near their parturition date, so providing them with a nest box and nesting material is a great way to support this natural behavior. We use the same box that you describe; we fill it with pine shavings. The box is given to the does a few days before their due date, and of course the mothers add their own fur to the shavings and make nice bunny nests.

Our nest box version is also made from stainless steel and has a removable floor, but

it not only has a front entrance but also a flip-top door so mom has privacy but we can peek into the box to check on her and the babies. I imagine something similar could be retrofitted from unused rat cages.

We also house the rabbits in stainless steel cages with plastic perforated flooring. In each cage, we insert a tray filled with sawdust for all rabbits and extra nesting material for pregnant females. It works very well. You can buy different types of nesting material that is certified, so this should not be a problem for GLP repro-tox studies.

At our facility, shortly before the expected birthing date, a HDPE [high-density polyethylene] hut is placed in the cage along with wood shavings, and the doe takes care of the rest. The HDPE plastic is nice, as it absorbs warmth and the rabbits don't bother to chew this hard material.

Protected social contact housing for male rabbits

Mature male rabbits are social animals, albeit quite intolerant of each other, yet it is common practice to cage them alone. Perhaps they would appreciate and benefit from protected social contact (e.g., perforated, grated or solid transparent cage dividing panel) with a neighboring male and option of visual seclusion (e.g., one half of the panel solid opaque)?

In my experience, bucks with this type of limited access are very aggressive and can injure each other quite badly through perforated cage dividing panels.

The risk of bite injuries associated with a perforated transparent panel could easily be avoided by (a) either reducing the diameter of the holes or (b) using solid yet transparent cage-dividers. With such an arrangement, would male rabbits ever rest side by side next to the divider, giving the impression that they enjoy each other's presence?

With the solid yet transparent dividers, we have observed increased vigilance behaviors, thumping, and charging the divider.

This was our experience as well. Even with only 2 inches [5 cm] of access, several males had scratches on their noses.

These observations are significant, in my opinion. I am wondering now, is there a consensus among those of you who have first-hand experience in this matter that mature male rabbits are better off caged alone—without any contact with another male—than with protected social contact options?

Our rabbit banks allow nose-touching at the top of the panel that separates two neighbors, but that's about it. I always arrange the banks across from one another so that the boys can see each other; when the banks are changed the tenants are moved to a different level, so they get a new view of different roommates. I have never noticed conspicuous aggression between eight males that were housed side by side.

I also have a six-boy situation with about eighteen girls kept in a separate bank but in the same room, and yet there are no signs of aggression between neighboring bucks while I am in the room. This is not to say that nothing is happening when I am not there, but if something does happen it is certainly not serious, as I never find evidence that injurious interactions have happened in my absence.

I took care of 12 single-caged New Zealand White (NZW) bucks who had access to puppy pens—clipped together for support—every other day. In these little floor pens, neighboring bucks were separated by bars that allowed limited physical contact. Mostly the contact was peaceful. A few boys would engage in spraying each other, but they never took it to any physical aggression. I spent a lot of time with these animals during their play time in these pens. Whenever I sensed something was getting out of hand between two neighbors, I'd either offer a distraction—a new toy or myself, which most of the animals seemed to enjoy—or if this did not do the trick, the two antagonists got a time-out and had to go back to their individual home cages for the day. I had to deal with only one serious ear injury in the course of several months, when a buck attacked his neighbor while I happened to be out of the room for a few minutes.

I also had one successful pairing of two bucks who gave me the impression that they could get along with each other; one of them was a bit shy, the other very relaxed and easy-going. One day I decided to allow them to have playtime together without separating bars. The two stayed so close during their playtime sessions that they looked like a two-headed rabbit! They spent the nights in their banks in separate cages, but they played together every other day in the little floor pen. Fortunately, this group of 12 rabbits was adoptable at the end of their study, and these

two boys went to a home together; to this day they are enjoying each other's company permanently!

Our group helped develop and test a double-wide cage with a special dividing panel. The front half of this panel is transparent with perforations allowing visual and minimal tactile contact between neighbors, while the back half is opaque and solid. Following a one-week acclimation, we monitored via remote video recording during a two-week test period the behavior of four pairs (eight animals) of male NZW rabbits in these refined cages, and eight male NZW rabbits in standard single-housing units.

We found a greater diversity of behaviors and an overall greater activity in the bucks who had protected contact with a neighbor. Much of this extra behavior was playful and exploratory. During the first few days, we had one buck who was thumping and charging his neighbor, but by the end of the week he had calmed down and stopped exhibiting these aggressive displays. The quasi-paired male rabbits spent a significantly greater portion of their time in the quadrant of the cage closest to their neighbor—resting or sleeping peacefully side by side in contact with the clear, perforated half of the dividing panel—than in any of the other three quadrants of their cage [Lofgren et al., 2010].

We've used these refined cages for close to three years now for both NZW and Dutch Belted rabbits without major injuries. We did have one unexplained split lip—this could have been a bite through one of the small contact holes; we are not sure, as nobody saw the incident.

Your encouraging findings dispel the misconception that adult male rabbits are only capable of negative or aggressive interactions with one another.

Based on my experience with macaques, I would assume that a male rabbit who gets along well with another neighboring male becomes socially more confident and will show less fear and aggressive self-defense reactions toward a human who approaches his cage than a rabbit who is always caged alone. Did you make any observations related to this (perhaps completely wrong) assumption?

We do find that the rabbits who have protected contact with a neighbor remain in the front half of the cage when an unfamiliar staff member enters the room and approaches their cage. They also have a shorter latency to touch when that person introduces her hand into the cage than single-caged rabbits. This does suggest that they are less fearful of people.

Acclimating rabbits to humans

Rabbits are biologically fearful of humans; this implies that being approached by and, even worse, being scruffed by a human is likely to stress a rabbit, hence influence research data collected from the animal.

Based on your own experience, what is the most practical and effective way of habituating rabbits to your presence and to being handled by you, without eliciting undue fear/stress responses?

When we get a new group of does, I will regularly spend some time sitting among them in their pen, gently talking to them and letting them approach me on their own terms; there is no hurry at all! I let them sniff and climb over me but will touch them only after they have made the first step and contact my hand. I don't think they will ever lose all fear of me, but they certainly tolerate me; this may be as good as it can get.

New rabbits—both does and bucks—in single cages seem a bit quicker to warm up. I also visit them frequently and talk to them and work my way up to petting each animal daily while I give them hay. When the rabbits come to the front of the cage after I have entered their room, I feel that I have accomplished a lot in terms of acclimating them to a human.

I believe that rabbits can become less fearful but not totally fearless toward people. This is not really surprising considering the fact that humans are their natural predators. Their fear reaction toward humans is probably a deep-rooted instinct.

Our rabbits LOOOOVE Fruity Gems (dried pineapple and papaya). When we get new animals, I give them these treats initially in their food hoppers. Once they have found out how tasty they are, I offer these treats through the cage bars, then open the door and place the treats in my hands. Unfailingly, the rabbits will go for the treats and I can then pet them while they eat. Most of our bunnies are pretty friendly and seem to enjoy it when I gently stroke them. I'm pretty sure they still don't like being scruffed, but they are so used to me that they relax very quickly after this disturbing procedure and take a treat from my hand.

Like with most animals [including humans], food can go a long way with a rabbit who is unfamiliar with you. Any time the rabbit can associate your hand in her or his cage with something yummy, you will buy some goodwill.

I like to hang hay balls in our rabbits' cages. They are suspended from the cage ceilings and look like round metal baskets; you fill them with hay and the animals will have to stand up to retrieve the hay from the basket. After the first few fills, the rabbits will eagerly try to get the hay out while you are filling the basket. Eventually it becomes easy to gently pat their heads or stroke along their backs while they are busily foraging; in this situation they will not dodge my hands, in fact some of them give the impression that they like it when I stroke them. I do this little ceremony every day; it does take a few minutes but it pays off greatly in rabbits who are relaxed when you touch them and who don't panic when you pick them up for a procedure.

Our rabbits are individually housed, and I can imagine that the presence of an unknown intruder (member of staff) induces fear.

From the moment of arrival and for the rest of the acclimatization period, we get our rabbits as quickly as possible accustomed to our husbandry and procedure staff. From day one, staff announce their presence by knocking on the door before entering the animal room, and talk to the animals in order to habituate them to their presence and voice.

I always say "hey guys it's only me" upon entering the room, and all is calm even if some of the animals can't see me. I like to name my rabbits and do call them by their name whenever I visit or get in direct contact with one of them.

As part of our husbandry procedure, all of the cages are opened at the same time and left open during the presence of the husbandry staff in the room. This entices the rabbits to come to the front of the cage and explore a different dimension of their living quarters. Most of them will sit right at the front of their cages, looking out into the animal room which has now become part of their cage environment, maybe even feeling less trapped by the physical limits of their cages but also realizing that their cage can be a comforting

retreat. We've never had rabbits jump out, maybe because they know how relatively safe their cages are.

As the days go by, entering the room induces less and less fear behavior—mainly frantically running around the cage or going into hiding—and more and more rabbits are sitting at the front of their cages, either with the cage front open or closed. The attending staff will also initiate non-invasive handling procedures, such as picking a rabbit up and performing daily health checks, in order to help the animals overcome their fear of humans and become more comfortable in their presence.

Our rabbits, once acclimated to humans, will readily come close to anyone who enters the room and calmly approaches their cages. Some of our rabbits have become so friendly that, when you open the cage door, they are right there, almost coming out at you in a welcoming manner.

We don't allow lab coats in our facilities so everyone wears disposable gowns over either their work scrubs (animal care/vets) or street clothes (research) so we [and the rabbits] don't have an issue with uniforms or what the person approaching is wearing.

My first defense against undue fear in my rabbits has always been background radio music/talk at a low volume. I think my rabbits developed a preference for a particular station on the radio. It buffers some of the husbandry-related noises that tend to upset the animals.

I never, ever enter a rabbit room without knocking on the door first; this avoids that the animals get startled when I come in.

RABBITS 69

My rabbits have learned through experience to associate white lab coats with disturbing situations; for this reason I wear blue or green scrubs or disposable gowns, to help them remember that they don't need to be afraid of me. When I am with the rabbits, I talk to them all the time in a calm and low voice. I weigh and brush my rabbits weekly and trim their nails; I give them no reason to be afraid of me. I like to let the rabbits initiate interactions with me wherever possible, let them approach, nudge or chin me as they wish. They are always first gently stroked before being picked up or being scruffed. I find that those, who are still making a fuss when I want to pick them up or scruff them, get less excited when I wrap them gently but firmly in a towel and then hold them against my body.

It seems to be very difficult, if not impossible, for rabbits to overcome their instinctive aversion to being picked up or scruffed. This is not really surprising because the natural raptors of rabbits do just that: swiftly lifting their prey off the ground.

My house rabbit loves being stroked on his head; he sits down, relaxes, closes his eyes and rests his head on the ground to lap up the attention, but he HATES being picked up. When being picked up, he breathes rapidly and struggles if given a chance—which could be interpreted as fear responses—but when he is put down on the ground, he makes no attempt to run away but just looks a bit annoyed, sometimes demonstrating his disapproval with a bunny hop and a vigorous flick of his back legs. To date he has not habituated to being picked up.

We recently received a batch of Dutch Belted males who were VERY fearful—some of them to the point of lunging at the handler and/or frantically trying to get out of the cage. I was asked to take a look and see what could be done.

After about an hour of visiting and talking to 10 single-caged rabbits, six were taking treats from my hand, two were taking treats using a dumbbell to serve them, and two were refusing any food—these were the worst ones, but I finally managed to at least open the cage door without them jumping out. The rabbits have settled down and there is no longer any stomping going on when a person is in the room. Today the investigator visited them and was amazed at the change in their demeanor!

In my experience, offering a treat every time a person approaches the cage is the best way to help a rabbit get used to the new living quarters and overcome his fear of people. Yogurt drops, fruit chewies, and hay are favorites that will quickly be associated with any friendly person who offers them. In order to make the animals feel relatively at ease when humans are present, it is very important

that not only attending care personnel but also the investigator visits the rabbits and offers them treats on days when no experimental procedures are being done with them.

Oral dosing of rabbits

Gavage of rabbits can be quite a risky procedure, especially when the animal is not sedated. In your experience, how can this risk be minimized, perhaps even avoided?

We avoid gavaging as much as possible. It is my experience that most rabbits will accept and swallow a drug if they have been properly conditioned prior to the actual study. I offer our rabbits a substance that they really like, such as baby food or pineapple juice, in a syringe. Once they get the taste and associate it with the voluntary syringe feeding, I mix the tasty treat with the actual drug and start dosing the animals for a given study.

Yes, that's the way to do business with animals! Marr et al. (1993) describe a very similar method: "We coated the tip of the syringe with sucrose. Inserting the syringe through the bars of the cage, we placed it in the animal's mouth and injected the sucrose solution slowly to allow the rabbit to taste and drink the fluid. We repeated the procedure three times a day for a total of 15 minutes per session, and within two days, 80% of the [10] animals voluntarily swallowed the fluid from the syringe. The [2] rabbits that did not seek out the syringe usually took it with only minimal encouragement.

At the onset of the therapy, we substituted the antibiotic for the sucrose solution. … We continued coating the tip of the syringe with sucrose granules throughout the therapy, apparently masking any unpleasant sensations produced by the antibiotic." The cooperative rabbits "would stand with their paws on the front of the cages, protrude their faces from between the bars, and appear to beg for the syringe containing the antibiotic [documented with a photo]." This non-stress method of "giving tosufloxacin was successful in producing the desired serum and bone concentrations."

If there are circumstances that really *necessitate* oral gavage, it should be possible to condition the rabbits with gentle firmness to allow one person to carefully insert the tube and administer the drug without stressing or harming them.

I have gavaged rabbits lots of times over the past five years and have never had any issues or problems with doing it. Our animal care technicians are really good with handling the rabbits and getting them used to being touched, held and restrained.

We simply make a "rabbit burrito" when gavaging. We tuck the rabbit up nice and comfy in a lab coat; the restraining person pulls the animal close to her/his body and the dosing person lifts the rabbit's head slightly up and forward—I arch my thumb and forefinger around the rabbit's muzzle and calm the animal by gently covering her or his eyes with my other three fingers—and simply slide the gavage tube into the esophagus and administer the drug—no sweat! If you go down the wrong tube, the rabbit lets you know immediately by throwing the ears forward; no reason for panicking, you just back out and try again.

In my experience, this procedure has

always been uneventful and easy-going. Actually, it never occurred to me that gavaging a rabbit could be risky. We have good people, gently and firmly holding the rabbits and skilled, compassionate people dosing them. We provide the rabbits and ourselves a non-stress, relaxed environment as much as possible. Sometimes we even have soft background music playing just for fostering a pleasant ambiance for everybody involved in the gavaging procedure.

Excellent! So, there are several ways of dosing rabbits without causing avoidable stress and possible injury. Seems to me that gentle firmness, patience and a few grains of compassion make all the difference for rabbits when we have to treat them, in this case by administering drugs orally.

We have had a few minor problems when gavaging our rabbits and learned from them. In the past, the rabbit would be held on her or his back and be handled by a single person who would insert the tubing behind the rabbit's incisors over the tongue into the esophagus. This method had been applied uneventfully for several years, but it was obvious that it was quite disturbing for the animals.

Now we are doing the procedure with two people and hold the rabbit in an upward position slightly turned toward the gavaging person. The other person carefully but firmly holds the rabbit by the scruff and presses the animal with the upper arm against her or his body. The gavaging person slightly cups the rabbit's lower jaw and nose region, inserts the feeding tube and administers the drug. I should perhaps add that as an additional refinement, we made it a strict rule that oral dosing does not coincide with any husbandry activities that are noisy and possibly disturbing to the personnel doing the gavaging and the rabbit being treated.

With this new method the rabbits are less stressed and only rarely show aversive reactions during the procedure, which is accomplished much more swiftly by two people than by one person only.

Recognizing pain in rabbits

I would like to draw on the group's wisdom regarding rabbit behavior. Specifically, what subtle signs tip you off to pain being experienced by a rabbit?

I don't have a lot of wisdom in this area, but one thing I do is offer rabbits Cheerios as a treat during health checks. This does not give a lot of information the first week the rabbits are in the facility, as they are still adjusting to their new environment; however, it is very useful once the rabbits are comfortable with their surroundings. If there is a rabbit who always comes to the front of the cage for a treat, and one day suddenly does not come forward, I know this rabbit does not feel well at all and needs immediate medical attention.

It is easy to pick out a rabbit who is off if I know how the animal behaves and responds to my presence when she or he is not in pain. When I have taken the time to get to know the rabbit individually under non-pain conditions, even subtle deviations of his or her behavior—especially decreased alertness—tell me that the animal doesn't feel well and possibly is in pain.

PRIMATES

Access to the arboreal dimension for monkeys

In their natural habitats, monkeys spend more than half of the 24-hour day above ground level in trees, on cliffs and on other elevated areas, out of reach of ground predators such as humans. How do we address the biologically inherent propensity of indoor-caged macaques to seek access to the quasi-safe arboreal dimension of their living quarters?

We recently revisited the space policy for our rhesus and stump-tailed macaques to better address the animals' need for free access to the arboreal dimension of their living space. We essentially doubled the height requirements outlined in the federal Animal Welfare Regulations so that all monkeys have proper access to the vertical space of their cages. All our cages are furnished with perches, and we've also started using nylon hammocks. The younger animals definitely take to the hammocks quicker than the adults,

but we do have plenty of adults, including aged ones, who use them as well.

Most of our old cages have been replaced by new ones that are a few inches higher, to provide sufficient room for the installation of a PVC pipe with a 2-inch [5 cm] diameter in each cage at a level that allows a grown-up rhesus macaque to comfortably sit on this perch without touching the ceiling and to move freely under the perch without touching it. The animals spend a lot of the time using their perches as lookouts, safe [and dry] resting sites above ground level, and a place to retreat during alarming situations. I think a properly placed perch should be a standard furniture of every monkey cage because it fosters the confined animal's sense of security.

Do caged macaques spend the night up on their elevated perch or platform?

It has been my experience when visiting caged rhesus macaques during nights that the animals were typically huddling on the floor even though they had access to comfortable PVC perches.

I have also seen caged rhesus pairs/singletons huddled on the cage floor during the night, oftentimes leaning against the provided perch.

During my evening checks past lights-out, most rhesus monks are sleeping on the cage floor.

Our caged cynos have several sleeping options. They can retreat to a comfortable hammock, a swing, a PVC pipe or a flat PC [polycarbonate] ledge. At night, they all seem

to prefer the flat ledge, which is the highest resting surface in their living quarters.

We frequently observe caged animals for 24 hours using a remote camera system at our facility. Our cynos are pair-housed; during the night, companions always sit together on a perch, never on the floor.

We have also used video cameras for overnight observations at our facility. Our paired cynos also always sleep huddled together on the perches in their cage. Our perches consist of three parallel, ~3/4-inch-diameter bars with a combined width of about 5 inches mounted from the front to the back of the cage. I have occasionally seen the animals leaning on the side or back wall of the cage while sleeping, but most of the time they are leaning on each other. I have never observed our monkeys sleeping on the cage floor. Our cages are also slightly unique in that we do not confine animals restrictively on the bottom quad. All animals have full vertical access permanently. And even though there are identical perches in the bottom half, I have never observed them sleeping on them.

I am wondering if we are dealing here with a species difference: in the three observations of rhesus macaques, the animals did not typically spend the night up on perches but on the floor while the cynos of the other three observations typically spent the night up on elevated structures away from the floor.

It could be that cynomolgus macaques have a biologically stronger need to have access to the arboreal dimension of their living space than rhesus macaques and hence spend the night up on perches, while rhesus macaques prefer to spend the night on the more stable floor rather than on elevated perches that may require some balancing maneuvers during sleep. This assumption is supported by the fact that in their natural habitat, cynomolgus macaques spend considerably more time up in trees than rhesus macaques do (Wheatley, 1999; Chopra et al., 1992).

Play cages/areas for monkeys

Does anyone remove their primates from standard caging and give them time in play areas? Do you singly house them in these areas or do you let them play with others?

I am using one large play cage for our rhesus macaques at the moment. It accommodates two pairs separated by a mesh wall. The cage is so tall that I can stand in it; it is furnished with swings and several perches at different levels. Pairs are transferred from their home cages to the play cage on Monday, where they get to stay until Thursday. I generally choose monks who have just come off study, so that they get a nice vacation.

I have never placed two single-housed guys in the play cage because of fear that they would start fighting, and I would not be able to separate them quickly enough to avoid serious injuries.

We also have a play area, which our cynos love! We simply had a fence company come in and construct two large pens in one of the animal rooms that was no longer in use. The pens are furnished with platforms, perches at different heights and toys. The monkeys get their vacations in these play pens either in pairs or singly depending on how they are housed normally.

When I first entered this field about 10 years ago in a university setting, I worked with a small colony of rhesus and cynos. All animals were singly housed but would be rotated, one at a time, into a large quad-style cage for the day. Although it looked great to us humans, the single-housed animal didn't use the space. Even with novel enrichment items, the animal was content to be able to perch higher than he/she normally could do.

My preference would be to house our macaques in compatible groups in large pens—automatically functioning both as play and exercise areas— instead of transferring them individually to a separate play cage/pen.

We only have 30 monkeys (rhesus) so we are able to do some neat stuff. We have floor-to-ceiling pens to which the home cages of the animals are attached. The monkeys have been trained to shift from the play pen back to their home cages.

Woodchip bedding for indoor-housed macaques

It has been repeatedly documented that woodchip bedding is a great way to foster foraging behavior in macaques while distracting the animals at the same time from stereotypical activities and aggressive interactions (Chamove & Anderson, 1979; Anderson & Chamove, 1984; Bryant et al., 1988; Boccia, 1989; Andrews et al. 2012).

How practicable is it to provision (a) indoor-housed macaques living in pens, and (b) indoor-housed macaques living in cages with woodchip bedding into which whole or crushed biscuits and/or seeds or other small foot items are scattered?

For indoor-housed macaques in pens this is definitely practical. Our cages, however, have open-grid floors that do not allow for scattering a substrate on the floor. We have tried installing shelves or small pans filled with woodchips so that the animals can forage, but our success was limited.

I think woodchip litter is fine for monkeys in pens. For monkeys in cages we haven't been able to also provide woodchips, as the cages are hosed out twice daily and the big worry is the floor drains getting clogged. I wish I could provide wood shavings to all my monkeys, but it would be quite a challenge to properly clean the woodchips out, as there is very limited room between the cage floor and the drop pans. As a compromise I throw forage crumbles or sunflower seeds, popcorn or raisins under the cages into the drop pans. The monkeys can reach through the grid floors and retrieve the various food items; these are so small that they can be hosed away during cage cleaning without risk of clogging the drains.

I worked at a laboratory that had bedding-catch-pans under the cages. We used woodchips and changed the pans three times per week. To promote foraging activities we'd scatter sunflower seeds, peanuts, and very occasionally meal worms, into the pan. The cynos definitely enjoyed spending much time picking out the treats.

When we had macaques housed in group-rooms, the floors were covered with shredded wood bedding; two or three times per day,

an access window would be opened and we'd toss the animals' feed ration, seed and cereal mixture, pasta, and other treats on the bedding. The animals spent a good bit of time throughout the day foraging.

When I was at another facility, we ran the individually-housed-macaque rooms dry. The drop pans were furnished with wood shavings mixed with forage (corn, sunflower seeds and other small food items). They were spot-cleaned daily and changed two to three times each week.

The indoor/outdoor pens were also spot-cleaned daily and forage was scattered on the fresh wood shavings. The pens were cleaned out and disinfected weekly. Once the pens were dry, the forage was scattered on the floor and an intact bale of wood shavings, wrapped in brown paper, placed on top of it. The monks knew exactly what was under the bale and would scatter the shavings themselves. It was fun to watch them go at it. Cleaning was labor intensive, just like mucking out stables, but the monks loved their foraging substrate. Disposal? We were lucky enough to have a huge EPA-approved incinerator.

Foraging and feeding enrichment for monkeys

There are quite a number of commercial foraging devices for monkeys on the market; some of them are excellent while others are of little use.

Please share your own experience with such gadgets. Which ones are effective? Do you bait them with standard food or with extra treats? How practicable are they in terms of loading and cleaning?

The first thing that comes to mind is that the forage boards, consisting of a piece of plastic with holes drilled in them—about big enough for a raisin or two—are not very effective. The primates get excited when the boards get filled with treats such as oats, nuts and seeds, but they are quickly cleaned out; I think these boards provide little in the way of actual foraging. It's more like a collection of miniature food dishes, and not much of a challenge. Puzzle Balls baited with frozen fruit—unfrozen fruit gets foraged and eaten up too quickly—are more challenging for the animals, allowing them to really use their foraging skills to retrieve the food.

Most of our non-human primates (macaques) like the puzzle boards that attach to the front of their cages. They spend a reasonable amount of time fishing the PRIMA-Treats through the small holes to the larger access opening. Our boys also like Primate Tubes with peanuts in them, Challenger Balls with PRIMA-Treats, and fleece/turf foraging boards on the outside of their cages. We usually smear honey or peanut butter on the fleece and then sprinkle foraging crumbles, Grape-Nuts cereal, or some other foraging mix on top.

None of these gadgets are too hard to clean, but preparation does take a lot of time, especially if you have a large colony. It can be labor intensive and time consuming, which is one of the reasons we still use the treat-baited Nylabone balls and Booda Yapples. If time is tight, they are quick and simple to provide; however, these items are nothing more than chew toys that give several minutes of distraction and a few extra calories. Kongs are good when they are filled with fruit and juice and then frozen, otherwise they do not draw sustained attention either.

My all time favorite foraging device is the Puzzle-Feeder. It is attached to the outside of the cage, is durable and is easy to clean. The path/maze configuration can be changed, thereby creating new challenges for the foraging subject; it requires only one small food item like a grape or peanut to keep an animal busy trying to retrieve it.

It is a shame that the Puzzle-Feeder has not yet been tested (I am not aware of any published findings) as a feeding device of the standard biscuit/chow ration. If the puzzle could be used to have the animals actually forage for their daily biscuit ration, it would provide sustained feeding enrichment without any extra time investment, unless the loading of the feeder is complicated.

As a general design, the Puzzle-Feeder is great for small items that fit through it, and that can promote long periods of foraging. I don't see a way that daily food rations could be used unless biscuits were made smaller, either from the feed vendor or by crushing them. Unfortunately, as they age, the gates of the Puzzle-Feeder start to break off inside the tab holes. The cost of these puzzles is prohibitive, so we had to stop using them.

I have tried several commercial foragers with our rhesus macaques. Most of these devices require the use of supplemental treats simply because they're not designed for biscuits.

Turf Foraging Boards are well received by the guys when baited with just about anything, for example, crumbled biscuits, peanut butter or seeds. However, as easy as they are to set up, they can be a real pain to clean if you use anything sticky—unless you get out the power washer. Astro Tubes are pretty much the same deal, but I have found that these are even better received by the monks because they have the spin feature. The tubes are easier to clean than the boards because I can stand them upright, and the downward flow of the water seems to work better than on a flat piece of turf.

Fleece Boards are great. Really easy to set up and to clean, as the fleece can be thrown away after use. But, because a monkey can directly pick up the food particles, a *good* forager can clean it off in no time.

Challenger Balls baited with PRIMA-Treats are also okay, but I don't use them very often at my current institution, as a majority of the animals are relatively old and some of them are missing fingers or have reduced fine motor skills as a result of research studies conducted with them. These animals get frustrated with the balls because they demand manipulative skills they no longer have. However, for younger animals with full motor skills, the balls are a hit; they are easy to fill and clean, but most animals can empty them so fast that you can hardly categorize them as entertaining foraging gadgets. For these animals, I replace the PRIMA-Treats with other more difficult-to-retrieve treats. A Challenger Ball filled with ordinary marshmallows works just great—drives 'em wild! The problem is, that it's very time consuming—bring your patience!—to load them. Not to mention that the clean-up can be a real challenge—for you, not the monkeys!

I've never met a monkey who doesn't like a Kong filled with some tasty, gluey stuff. These rubber toys have the disadvantage that it can take quite some time to prepare them, and I have found they are a real nightmare to clean, even with a test-tube brush. The Booda Yapples get a similar response from the monks as the Kongs, but they are much easier to clean.

The monkeys really like the rubber, treat-dispensing Mike Toys, which are designed and marketed for dogs. They are relatively long and narrow, so it can be a little tricky to fill and clean them. I have discovered, though, that monkey biscuits fit in the larger ones quite nicely. So, if I have an animal who is not supposed to get treats, I use this toy because I can bait it with their standard diet.

I've used a variety of commercial PVC feeders, none of which I found to be useful. Some of the hanging types are designed for biscuits, but the biscuits don't move well, which typically frustrates the monkeys quite a bit. Rather than patiently try to move the

biscuit forward, they simply rip the feeder from the cage front and then shake the biscuit to the bottom and pull it out. I don't want to frustrate my monkeys too much, so I've switched to using peanuts, but the animals then empty the feeders faster than I can fill them. Not a very satisfactory solution!

The Crumble Disk Holder is normally well accepted by the monkeys. Super easy to set up and clean unless the disks sit for a while and turn into paste on the inside of the feeder. If the disks sit, I will not give the gadget back to that particular monkey.

As cool as it looks, the Shake-A-Treat only seems to frustrate the monkeys, and it's a nightmare to clean this device unless you have the correct tools to disassemble the beast.

The Puzzle Toss is great for fruits and biscuits of all types, but every monkey I have ever seen using it managed to get it apart in a short time; thus it loses its oomph, so to speak.

My personal favorite of all commercial feeders specifically designed for non-human primates is the Universal Bracket. It's an adjustable bracket that can hold different fun items such as turf boards and tube feeders. It's a bit time consuming to prep if you have a large colony, but it's a breeze to put up, and the boards clean very easily especially if your facility is fortunate enough to have a dishwasher.

We've been using Kongs, E-Balls, and Turf Foraging Boards for our rhesus macaques. I'm not impressed with any of them, for a number of reasons: all require a considerable amount of time to load and to clean, and all are fairly expensive.

Altogether, I have found self-made foraging devices to be of greater use than the commercial ones.

Even though some of the commercial feeding/foraging devices are real hits for the monkeys, the time investment for loading and cleaning the devices make it problematic for institutions with large numbers of monkeys to implement them as standard foraging enrichment. When you take care of 100 macaques, which includes feeding and cleaning, blood collections, TB testing and holding animals for treatments/examinations, the time for daily extras becomes very, very limited. Using the daily biscuit ration in such a way that the animals have to forage, i.e., work to retrieve the biscuits one by one, would be an ideal foraging enrichment option for such a situation.

I agree, it is very helpful whenever the standard food ration can be utilized for foraging enrichment!

Now that we have discussed a few commercial foraging devices, have any of you developed and tested custom-made feeders that promote foraging behaviors/activities?

Really cheap and simple are 4-inch-long pieces of PVC pipe. I smear a bit of peanut butter or yogurt on the inside of the pipes and put them in the freezer for a few hours. Our macaques seem to have a great time retrieving the tasty stuff from the pipes with their little hands. It can be a bit messy, but who cares!

I know an enrichment technician who presented a poster on the use of self-made puzzles for feeding the daily biscuit ration at a National AALAS meeting. He told me that they now feed all their rhesus and baboons with this new device.

[Here is an annotation of this presentation: The feeder dispenses monkey chow and fits on non-human primate group four quad rack cages … . The original feeders dispensed 18 to 20 biscuits. At feeding time, the macaques removed all the biscuits within 3 min, and those that were not eaten or stored in cheek pouches were pushed back through

the feeder onto the room floor or dropped through the cage floor grid. ... Each feeder took approximately 1 hour to make and costs approximately $60 in materials ... . Puzzle feeder implementation increased time spent foraging (approximately 20 min per biscuit), reduced food wastage, and decreased clean-up time (Glenn & Watson, 2007).]

In order to make feeding enrichment practicable for each animal in a large rhesus and a smaller stump-tailed macaque colony, I used structural elements of the cage and turned them into food puzzles. Without extra costs, I (1) moved the standard food boxes a few inches away from the large access holes, or (2) distributed the daily biscuit ration directly on the cage ceiling.

In both set-ups the animals had to use skillful foraging techniques to retrieve their daily biscuits through the mesh of the front or ceiling of the cage. This simple refinement resulted in a many-fold increase in the time that the animals spent retrieving their daily food ration; it also decreased food wastage because the animals ate all the biscuits that they had laboriously retrieved. Working for their standard food, rather than collecting it without effort, did not affect the macaques' body weight maintenance (Reinhardt, 1993a,b,c).

The mesh of the cage floor also provides a kind of food puzzle. I distribute small treats such as mini marshmallows or Fruit Gems on a sheet of paper placed on the cleaned drop pans of marmosets. The animals have to reach through the mesh, try to get hold of a treat and

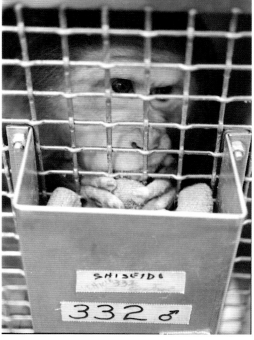

PRIMATES 81

retrieve it. This foraging activity keeps them quite busy.

Recycled glove boxes stuffed with shredded paper towel that contains mini marshmallows or other small treats provide effective, yet inexpensive foraging gadgets for our mamosets.

Simple bird suet feeder baskets are commercially available. They turned out to be my favorite foraging devices for our rhesus macaques. I buy the baskets in bulk and use them for everything from biscuits to fruit, to frozen blocks of juice. It's even good for guys with limited motor ability—I just have to hand them the basket rather than hang it on the cage front.

Suet feeders are inexpensive but do provide suitable foraging enrichment for our macaques, as well. We usually hang them on the inside of the caging so that the monkeys can manipulate them as they wish; yes they chew them up and make them quite raggedy in a short time. When they get too bad, we toss the little baskets and replace them with new ones.

One thing to be careful of when buying suet feeders is the size of the square holes of the grated baskets. They must be larger than a monkey finger (so that the animal can use a finger to reach the contents of the feeder) and smaller than a monkey's hand (so that the hand cannot get stuck in the device).

Lastly, the monkeys can easily open the top of the suet feeder basket and empty it very quickly without actually foraging. To avoid this, I close the top of the baskets with a small zip tie.

We load the suet feeder baskets with soggy nuts/seeds/biscuit/fruit slurry and then freeze them. The frozen blocks are nice because they take a few hours to melt, giving a gradual foraging-type experience for the monkeys.

Is it practicable to offer caged monkeys corn popped in the animals' room?

This is one of my favorite types of enrichment. It is so entertaining and fun not only for the animals but also for me! Popcorn, unlike most other treats, is low in calories, which is pretty cool.

I do this for our rhesus and cynos about twice a month. They have their own air-popper that never leaves the area. I use large plastic tubs to catch the popped corn. The monkeys seem to like watching the popcorn emerge, and certainly enjoy the aroma.

 I will sometimes throw the popcorn over the cage tops to let it "snow," or go around and let everyone get a couple handfuls directly from the filled tub, or I might put it on a paper towel on top of the cage. Of course, I may just pass it out; popcorn is something everyone will take from a hand. I think it is a useful human-monkey bonding tool.

 My favorite thing about the air-popped corn is its low calorie, not-junk aspect. I am always trying to provide the animals the healthiest extra food possible.

Our monks get popcorn handed out right after cage-cleaning several days a week. They seem to love it, and the personnel get a chance to foster a positive relationship with the animals.

Our cynos LOVE popcorn. I pop the corn right in front of them. They always get so excited when they see the popcorn machine! Sometimes popcorn will fly out in unpredictable patterns, to the great delight of the monkeys. I mix raisins or nuts with the popcorn and let them grab their share directly from the bucket. Some of the monks are greedy, of course, and take several handfuls, but then you also have those few who are very picky, sorting through the bucket at great length until they finally find that perfect piece of popcorn; it's very cute! Of course I have to play movies while they are eating their popcorn, too! It's a fun time not only for them but also for me. I always enjoy seeing them get so excited and happy.

When I was working with our cynos I popped corn in their room twice a month on Fridays—kind of a Happy Friday for all. They loved it

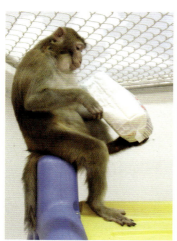

of course; the sounds, smells, and especially the corn itself popping into the container got them so excited. On these days we would also play music and I would set up large rotating disco balls that would send colored dots all around the room. Our cynos would try so hard to capture the dots as they moved across their cages. Like everyone else we were short-staffed, otherwise we would have organized these Happy Fridays more frequently. It was very rewarding to see how excited the animals would become once they saw the popper enter the room on the enrichment cart we had set up.

I've introduced the hot air-popper here at our facility also. It's the best thing ever. I love seeing the monkeys' [cynos] faces as they smell and watch the corn pop up and out of the machine; then best of all they get to eat it. We bring the popper in the animals' rooms once a week on the day when they also get a TV; it's awesome.

We offer popped corn in the room several times a week, especially in the cold winter months. Our cynos and rhesus love it and don't seem to tire of it, no matter how often we provide it. We snack on it as well—note, we never eat *in front* of the monkeys, but we do eat *with* the monkeys!

We offer popcorn to our macaques regularly. Corn is popped in the anteroom with the door open, so the monkeys hear it's coming.

Popping corn in the animals' room does provide enrichment in which neither the monkeys nor the attending personnel lose interest over time. It's easy to provide plus it

doesn't cost much—the perfect environmental enrichment!

All the monkeys on our campus receive air-popped corn at least once a month. However, we have monkeys of different species (vervets, rhesus and cynos) spread between rooms/buildings. Currently we pop the corn in a non-animal room; we are not sure if it could create a cross-contamination hazard when the same popper is moved into rooms of different primate species.

I wouldn't see a problem, using the popper in the animals' rooms even though monkeys of different species are housed in them. After all, the monkeys are not climbing over the popper, so they have no contact with it other than the popped corn that you are handing out. And even if some over-cautious superior says *no*, why not purchase a few poppers, one for each species? These gadgets are not expensive but they pay off quickly in both the animals and you enjoying a few fun moments each month in which you and your little machine become a highlight for the monkeys.

I am looking for foraging devices that (a) can hold a trail mix (consisting of nuts, seeds, grain), fresh fruits, and veggies, (b) are easy to fill, (c) are challenging for the animal, and (d) can be washed easily. Can anybody please share suggestions based on first-hand experience?

At our facility, foraging items such as trail mix, fresh fruit, frozen fruit, and vegetables are directly distributed on platforms. This allows the *fascicularis* macaques [cynos]

to engage in species-appropriate—albeit basic—foraging activities. We offer this foraging enrichment every day, without spending much extra time for distributing the food stuff and cleaning the platform.

We use various commercial feeding devices constructed from PVC. They are usually filled with grain, treats, or sticky substances (e.g., peanut butter) and hung on the outside of the primate's cage. You have to invest a bit of time to fill these devices, but the cleaning is easy: we soak them in water and bleach for 20–30 minutes and rinse.

Here is a neat device we are using, and that you can easily make yourself:

1. Cut numerous holes in a 12-inch [30.5 cm] length of PVC tube with a diameter of 2–2.5 inches [5–6.4 cm]. The holes can be cut in various sizes to accommodate whatever food items you intend to put into the tube.
2. Add a screw cap to the bottom and another screw cap to the top of tube. The screw cap on the top needs to have a hole drilled in the center. Through that hole, put a stainless steel eye bolt with a lock nut secured tightly on the underside; make sure the hole is a tad bigger than your eye bolt.
3. Hang the device on the outside—or inside—of the cage with a quick link through the eyebolt at the top; because the hole is a tad bigger than the eye bolt, and the lock nut is past the threads on the bolt, it spins and makes it easy for the monkeys to access all of the holes.

Loading this device on a daily basis does not demand much time, and to soak it in bleach water followed by thorough rinsing is really not a big undertaking. It's certainly worth it when you see how your animals take the opportunity you are offering them to engage in skillful foraging activities every day.

Fruits and vegetables cannot be categorized as foraging devices but they do have a similar foraging enrichment effect. If presented whole, fruits and vegetables allow monkeys and apes to engage in natural food processing activities.

All our rhesus macaques [approximately 950 animals] receive every day—including weekends and holidays—one piece of produce, which may be half an apple or orange, one whole banana, one corn on the cob, one sweet potato, a generous section of a watermelon, or a generous section of a pumpkin, as an integral part of our environmental enrichment program. The monkeys get these supplements in the late afternoon after they have finished their daily biscuit ration. I don't remember a single case of an animal being adversely impacted by processing, enjoying and ingesting her or his daily fruit or vegetable. There is not much time required to prepare these foraging items and they are easily distributed. Fruits and vegetables don't need to be cleaned after usage, but some animals may leave a mess behind that needs a little extra attention when cleaning the cage.

Does anyone on the forum offer mangos or papayas to their macaques and/or vervets? I am wondering if it is safe to give whole mangos so that the animals can engage in food processing activities, gnawing and tearing through the leathery skin, eating the fruit off the large seed and, finally, gnawing at the big seed. Papayas are relatively large, so they would be cut and handed out in smaller pieces to the monkeys.

Our rhesus get mangos from time to time. I cut them up, as per this facility's protocol. The monkeys typically play with the seed—which is the size of a monkey fist or a bit larger—and finally gnaw it into small fragments. We have not encountered health or dental issues related to the mango seeds.

We have worked a deal with Costco where we pick up their expired or bruised produce once a week.

After we weed through the gross stuff, we usually end up with enough whole fruits to distribute to the 450 vervets in our colony, plus others monkeys on the campus. Our animals get lemons, limes, mangos, papayas, raspberries, blueberries, pummellos, grapes, watermelons, cantaloupes, tomatoes, tangerines, citrines, oranges, a variety of apples, kiwis, bananas, green beans, lettuces, Brussels sprouts, bell peppers, persimmons, and other fruits.

We have never had a problem but are aware that some monkeys may have a mild

reaction to the skin of the mango. They often carry the hard seeds around with them for one or two days and shred them into thin leather-like strips with which they play but do not ingest.

Initially, our monkeys were a little confused by the large numbers of small, glibbery seeds in the papayas, as they do look a little like fish eggs. It often takes vervets, in particular, a while to try something new.

The monkeys seem to really enjoy and explore the variety of produce offered. They have never refused to eat a fruit or vegetable in particular, though they do have their preferences.

Your animals are really lucky. By giving them such a variety of whole fruits and vegetables, you not only provide them foraging enrichment but also feeding enrichment. That's what we humans also enjoy: eating a variety of food items that differ in taste and texture.

Did you have some sort of formal paperwork to set this deal up with Costco?

We do not have a formal arrangement. We knew that Costco had started giving their expired produce to a pig farm, so we simply approached the produce department manager to see if we could also get expired produce for our monkeys. Once a week we go to the loading dock, ring the bell, and they pull out the giant bin for us to dig through and load up in our car. The pigs still get plenty too.

I was wondering if the monkey folks would care to give me a run-down on produce portion sizes for your rhesus and cynos.

Our facility manager has greatly decreased the amount we are permitted to offer, and we would like to make sure that we hadn't been over-doing it. She also asked for any documentation from others about what they believe to be standard.

Prior to the restriction, we were giving all rhesus, for example, a quarter or half apple, orange or banana, plus one leaf of lettuce, a half stalk of broccoli or a quarter pepper.

That's about what we do also. We distribute the produce at the end of the day as a kind of

PRIMATES 87

supplement after the animals have eaten their daily standard rations of chow.

We offer our cynos one piece of fruit or vegetable twice daily when they receive their portion of the daily biscuit ration. Depending on availability, the pieces of produce consist of:
> a quarter apple, pear or orange,
> a quarter grapefruit cut into half,
> a half or third of a banana,
> approximately 2.5-inch-long pieces of cucumber or carrot,
> a small handful of grapes, baby carrots or berries, or
> a half corncob in the husk.

We offer each of our rhesus, bonnets and cynos a half-cup serving of banana, apple or orange pieces at least four times a week. The produce is fed in the late afternoon so as not to interfere with consumption of the daily chow ration.

The daily produce supplement of our cynos may consist of:
> a half apple, banana or orange,
> one celery stalk,
> a quarter sweet potato, or
> a half carrot.

Can any members on the forum share experiences with feeding their monkeys frozen fruit or frozen fruit juice as a form of environmental enrichment and, if so, how did the animals respond? Were there any discernible adverse impacts on their health and/or food intake?

It is my experience that feeding frozen juice or drink mixes to macaques, squirrel monkeys or owl monkeys entertains the animals quite a bit without negatively impacting their health and standard food intake. My only recommendation is to NOT use red juices or red-colored drinks. You can really freak out your vet staff if they walk into the room and see red everywhere—on the floor, on monkey faces and in drop pans. Also some of the drink mixes will stain your floors, so be careful if you have any inspections coming up.

We bought a bunch of funky-shaped ice cube trays. They are silicon so they clean really well in the cage wash; the different shapes allow us to more easily make small sizes for our New World monks. A favorite recipe here is water with chopped cucumber in the middle.

Our rhesus and cynos get frozen treats very frequently; it's a favorite item here!

I use ice cube trays or paper cups and freeze Kool-Aid, juice, yogurt and applesauce mixed with fruit and veggie bits in them. I also freeze chunks of cantaloupe, watermelon, pineapple, banana, strawberry and other fruits of the season for our animals. All of them seem to enjoy the frozen stuff, which does not affect their health or well-being in any noticeable way. We haven't had anyone who wouldn't eat their regular feed ration due to frozen treats.

Our rhesus and cynos love crushed frozen Prang that I throw into their cages by the handfuls. They reach through the mesh floor and forage for the ice chips on the drop pan. Little paper cups filled with frozen orange juice or apple juice seem to make the monks very happy. I also load commercial foraging devices with chopped frozen fruits—including watermelon. We haven't had any health issues or eating problems related to the frozen juice and fruits that we give our animals on a regular basis.

Macaques also love normal ice cream!

We freeze Tang and Jell-O in Dixie cups and distribute those to our rhesus and cynos. Kool-Aid frozen in ice trays and frozen fruits are also favorite foraging enrichment items for our monkeys. We have never had any diarrhea or a sick animal due to these frozen enrichment items.

I give frozen, certified treats—such as Fruit Crunchies, Fruity Gems and Fruity Bits mixed with fruit and veggies in Dixie paper cups at least once a week to our cynos and rhesus. I haven't had any issues with the animals not liking these frozen treats or with health problems. If the study directors ask me not to offer their animals the treat in the cups, I just take it out and hand these animals only the frozen block, so they still get to enjoy the treats!

I have been hearing a lot about the increased nutrient content of produce in season, and therefore stress in our feeding guidelines that our monkeys should receive seasonal produce whenever possible. In trying to think of spring

PRIMATES 89

things, I wonder if anyone offers asparagus or rhubarb and whether your monkeys like this produce.

Our cynos and rhesus love both, but I have noticed that most animals like the asparagus better if we steam it. They also like the rhubarb, which they enjoy stripping into little strings before eating it; this is a great enrichment activity!

Our animals also get whole coconuts; they love grooming the nuts until all fibers are pulled and picked off the shell. Another fruit that our monkeys get is tamarind, which we buy in bulk. Tamarind looks like a huge brown pea-pod; it is relatively hard shelled on the outside. When you peel the hard stuff away, the inside has the texture of a fig, but tastes like a really sweet lemon. Our monkeys will literally do cartwheels for one of these! I like them too!

We also try to feed in-season fruits and veggies. Now in the spring season our macaques get asparagus (some rhesus really like it, others are not so thrilled by it), lettuce, peas, spinach, cabbage, radishes and strawberries.

I get huge organic watermelons and pumpkins for a very good price from a local farmer at the end of the growing seasons in late summer and early fall. Both the caged and the pen-housed rhesus monkeys can't get enough of them. I distribute slices of melons and pumpkins to the caged animals. Animals living in pens receive the whole fruits; they would first bite a hole in the thick and hard rinds and then dig with their fists into the interior flesh, retract their hands and lick them with gusto. It always takes at least an hour until the melon or pumpkin finally breaks open, allowing the monkeys to finish all of its tasty contents.

Through a national supplier, we can get a pretty good variety of fruits and veggies year round. I try to buy the more expensive items when they are in season, but sometimes it is nice to surprise the monks in the winter with treats like strawberries or blueberries.

We use national and local suppliers for seasonal produce, but we also have a small *primate garden* in which we grow some vegetables and herbs. We are currently awaiting fruit from our newly planted fig trees. Another thing that is *fruitful* (sorry, couldn't resist!) is calling around to grocery stores and churches for gourds and pumpkins around Halloween when they have a surplus that they are usually very willing to let you have in exchange for hauling them away.

It seems to me that feeding enrichment can enrich the often rather monotonous lives not only of monkeys but also of us who are taking care of the animals; I love the *primate garden* project.

Gnawing sticks for monkeys

Based on your own experience, do you recommend gnawing sticks/blocks as effective enrichment objects for monkeys, or have your observations shown that gnawing sticks/blocks are useless enrichment items?

We just gave our rhesus and cynos 12-inch-long branch segments of littleleaf linden (*Tilia cordata*) trees; the animals have a real "gnawfest," so obviously they like them.

I work with both rhesus and cynos. We offer both fresh wood and purchased wood. The monkeys definitely seem to prefer the fresh stuff and it can keep certain individuals busy for hours! We have one guy who is pretty adept at shaping the sticks into spears.

Most of the wood is claimed from trees on our facility's property. I also contacted our local zoo, which often has browse donated for its animals; I got permission to help myself when this happens. We have a pretty extensive list of vet-approved trees that we can make use of for our monkeys: alder, amaranths, aspen, bamboo, beech, birch, bush honeysuckle, butterfly bush, cattails, chicory, clover, cottonwood, daylily, dogwood, elaeagnus, elm,

fig, forsythia, grapevine, grasses, greenbriers, hackberry, hawthorn, hazelnut, hibiscus, Japanese silver grass, kerria, maple (except red), mock orange, mulberry, nasturtium, Oregon grape holly, pear, pickerelweed, poplar (except tulip poplar), raspberry and blackberry, redbud, rose, snowberry, violets, water hyacinth, and willow.

I have only used a few of these species so far, but the maple and grapevine are definitely the favorites so far. I just handed out willow branches the other day which were also a big hit!

We don't have drains in our rooms; the husbandry staff remove the chewed-up material as needed and replace the branches when they are soiled or worn down to small pieces of wood.

We used to have a lab technician who would cut 12-inch-long oak branches or other hardwood branches for us. We gave these to our baboons, rhesus, and cynos. The monkeys loved the oak because they would tear off the bark and eat it, and then use the rest for tactile enrichment. Unfortunately, the lab tech was unable to continue providing these gnawing sticks due to a medical issue, so we turned to bamboo. We have some wild bamboo growing out at our farm and we harvest it ourselves.

Anything we bring into our facility has to be autoclaved, so the bamboo leaves turn brown and crispy, which our monkeys don't seem to mind. The autoclaving of the bamboo makes the shoot part very brittle and it can, therefore, splinter easily. We are very careful to give out small pieces of bamboo that have a lot of leaves and less shoot.

All gnawing sticks are changed out completely every two weeks, or earlier if they look soiled or are worn down into small pieces.

I provisioned more than 700 pair-housed rhesus macaques with segments cut from dead red oak trees for more than 10 years. The animals used these natural wooden objects about 4% of the observed time for gnawing, manipulating and playing, without recognizable health hazards. There was no sign that long-term exposure to the regularly

replaced gnawing sticks diminished the animals' interest in them.

[Line & Morgan (1991) provisioned 12 single-caged adult rhesus each with a routinely replaced almond wood gnawing stick for an extended period of time; individuals actively used their stick about 6% of the observed time, without adverse health effects.]

We give our rhesus, pigtailed and cynomolgus macaques commercially prepared manzanita wood as a manipulable object in the cage or as an item hung from the front of the cage. It is well utilized, as shown by wear and through video observation. We use it in rotation with toys and enrichment devices made of other materials. We have been doing this kind of enrichment for at least eight years with no clinical or drain problems.

The manzanita wood is sanitized in the same way as other enrichment objects. Cage washers effectively sanitize it (Luchins et al., 2012).

When you have made use of gnawing sticks for macaques, has it ever happened that regulatory inspectors raised concerns regarding the cleanliness of the wood?

No. Right now we use either red oak or manzanita; when soiled, the wood is thrown out. We are just about to receive a large order of gnaw sticks for the colony.

We have been using gnawing sticks for our macaques for eight years and, so far, have never had any problems with them during inspections. The wood segments are cleaned daily during the cage cleaning process. I don't remember that we ever had to throw a gnawing stick away because it was unduly soiled, but we have to replace them regularly before they become so small that they can pass through the mesh of the cage floor.

Does anyone at a tox facility use wood/gnawing sticks?

We use wood in our tox facility for our monkeys. It is a hardwood, but I am not certain exactly what type. We give it out in the cages and also have them drilled so that they can hang outside, with foraging holes stuffed with small food items. We also use wood shavings in our large gang-housing cages.

Water as enrichment for monkeys

Providing monkeys with swimming pools during the hot summer months is probably a most attractive environmental enrichment for them. Macaques and baboons are good swimmers and divers, but just simply playing with water can fascinate them for extended periods of time.

If you provide the monkeys in your charge with suitable containers filled with water on a regular basis, what are the hygienic and sanitary implications?

Both are definitely of concern at my facility and we have staff lifeguards present for all pool times. Therefore, pools are not a normal form of enrichment due to the time investment. I was not privy to the conversations and concerns, but know they were pervasive enough to inhibit pools as rotational enrichment for our outdoor-housed rhesus macaques.

Your comment surprises me a bit.

Has anybody encountered problems with monkeys drowning in water or being harmed in any manner when they have access to water?

We had an incident a few years ago where a male cyno unexpectedly jumped on top of his female companion while she was under water, and appeared to be deliberately holding her under. That particular pool was only about 2.5 feet [76 cm] deep. Just as I jumped to go rescue her, the male let go. She shot straight up out of the water with anger in her eyes. As soon as he saw how angry she was, he ran

quickly to an adjacent attached enclosure; she chased him down, caught him, and punished him. He's never done this since, and I have never observed an incident like this with any of the others, but because of that one incident, all deep-swimming has to be monitored. We don't have to monitor running-hose-water and really, the monkeys seem to enjoy that as much as the swim time. Also, even though we monitor deep-water swimming, we always have something in the pool that the monkeys can use to easily crawl out of the water if needed. Right now we have a log—too heavy for the monkeys to move—that we placed diagonally in such a way that it extends from the bottom of the pool out of the water for several feet. From anywhere in the pool the monkeys can easily get on the log and just walk out (for the sissies).

Our indoor group-housed cynos get access to a 5 x 5 feet [1.5 x 1.5 m] kiddie pool several times a month. The water is 3 feet [1 m] deep. The animals pick up small floating treats from the surface of the water; they swim and dive for grapes and other fruit items. They like it very much. We exchange the water as needed and have encountered no hygienic problems; also, it has never happened that an animal got harmed in any manner in the pool.

We recently placed a large water-filled tub into the activity unit of one of our adult rhesus males. He took a couple of days to get used to having the tub in his cage. While he seemed very interested in splashing around, he didn't actually jump into the water. Also, the tub served as a big toilet for him. He found it incredibly amusing to pee and defecate into his tub of water from a high perch. Needless to say, within about three days the tub became a hygiene concern and that was the end of it.

At my last facility we used kiddie pools with our outdoor-housed cynos and rhesus. The cynos spent a lot more time in and around the pools. It was really cool to see the juvenile cynos actually swimming underwater with their eyes open! Some of the juvie rhesus would get in the water, but only stay for a few seconds and dash out again. Adults of both species would bob for treats in the pools and splash a bit but not really enter the pool.

PRIMATES 95

We also use kiddie pools. When the monkeys are finished playing, the pools can easily be dumped out and re-filled with fresh water.

Our cynos enjoy running water as much as swimming! Sometimes we'll take an old hose,

turn it on and slide it into their enclosure. They LOVE it! Needless to say, they will also trash the hose in a short time.

Protecting watering system hoses

I would love some input on a problem we are encountering. Our monkeys are having a field day with the automatic watering system. They can reach through and pull out the hose attached to their cage, thereby disrupting the watering system. Has anyone had to deal with this?

Ugh, we had this problem too. We pull the cages off the wall about 2 feet, just so hoses are taut enough that they can no longer be reached by the monkeys. It actually works nicely because then techs can walk behind the cages easily, checking and maintaining the water lines.

I have only had a couple monks do this. To stop them from creating a problem, we took zip ties and tightened the water hose down to the attachment so that they get out of the monkeys' reach.

Pair and group formation of monkeys

What tells you that two adult cynos have established a dominance-subordinance relationship during a non-contact familiarization period before you introduce them as a pair?

Cynos seem to have this *thing* where it just takes them a long time to really find out who is dominant and who is subordinate. It can go back and forth and back and forth for quite some time. I'd rather they had it thoroughly worked out before putting them together. This is what I have learned over the years:
1. If both partners are lip smacking and acting friendly all the time, THAT is a deal breaker and I wouldn't even attempt the pairing.
2. If both are acting dominant and are threatening each other, that is also a deal breaker.
3. If one or the other reacts strongly to room dynamics and exhibits a lot of redirected aggressive behaviors, again that is a deal breaker.
4. If it *appears* the two may do well together, we will put a food dish down

in front of each monkey, real close together in front of their caging, so that each partner—while still separated by the mesh panel—can see the other approaching and eating the food. What happens at the dinner table is critically important and can really give you some good additional clues as to what could most likely happen when you pair. The clues are sometimes very brief and could easily be missed, so you REALLY have to watch every second during this event; don't blink. You want to see that one partner goes directly to the dish without any hesitation and starts chowing down. The other monkey should NOT be so quick, but should cautiously glance toward the monkey eating, and sort of *ask for permission* to approach his or her own dish; I call this *being polite*, and this clearly distinguishes the subordinate from the dominant partner. Then when he or she starts eating we watch every gesture of the first monkey, making sure he or she is a *giving-permission* type, that means a dominant animal who accepts the presence of the subordinate companion.

There are always exceptions to the rule, but it is my experience that the above criteria are helpful when you want to end up with a happy, i.e., compatible pair.

Some of our cynos are severe human-abuse cases and they are so messed up emotionally that their behaviors are all over the place; this makes it even more difficult to determine a monkey's rank status and predict if he or she will match up in a compatible pair.

I am looking for some help on pair- and group-housing of adult female African green monkeys [vervets]. I find their behavior hard to read; they do not display a true rank relationship that I can recognize.

I briefly worked with 16 adult female vervets and did not have much success in pair-housing them. Cage companions were very sweet to one another when people were in the room. The moment they were alone, they fought. I don't know if my experience was typical or not, but I hope you have better success.

My experience is similar. I successfully paired two pre-familiarized adult female vervets, only to split them already on the third day when one of them was screaming after an injurious fight and the other merely walked past as if nothing had ever happened. I was unprepared for this because the remote video camera had shown that the two had spent most of the time amicably grooming each

PRIMATES 97

other on the first and second day after pair formation.

I have worked with African green monkeys (AGMs) for many years and have no problems observing and determining rank relationships, but there is indeed a minority of individuals who completely ignore each other, giving the impression that they have no rank relationship. Another problem is that once rank has been established, that does not necessarily stop the fighting. Relationships can be relatively unstable and break down after only a few months of compatibility, resulting in fighting and even injury. However, to place this all into perspective: we find that on average about 80–90% of adult female pairs formed from unfamiliar individuals remain compatible without short- or long-term problems. We accept a certain number of relationship failures and simply look for another partner. However, if you have only a few individuals you do not have that luxury. In our situation, we find pair-housing of females more attractive than group-housing, although we practice both, particularly when raising juveniles to adulthood.

When pairing or grouping AGMs, we have no magic formula or proven method; intense observation after pairing/grouping is obviously most important, and we do not pair or group in the afternoons and before weekends. We do not separate pairs at the first sign of trouble; it takes about two to three days until a new pair settles down, sometimes amidst much screaming and cackling.

We have tried familiarizing animals before pairing or grouping but found no difference in the outcome. We had partners who groomed each other lovingly for days

through a divider mesh, but started fighting with each other the moment this mesh was removed and they had full physical contact.

We have never succeeded in pair-housing unfamiliar adult male African green monkeys.

What is your experience with forming small groups of squirrel monkeys? Does anyone have experiences to share involving all-male groups of common and/or Bolivian squirrel monkeys?

We form and re-form new pairs and groups of more than 15 adult *saimiris* on a regular basis with no major problems: male-male, female-female and male-female pairs/groups. We always use a new cage for group formations and stand inside the cage during the first moments, just in case there is trouble.

The most important thing is to use a new enclosure when grouping to avoid "resident effects". Typically we group one or two individuals with a big stable group in a cage that is new for all animals. If one of the new animals is a male who causes too many or too violent fights, we typically keep him in a cat transport cage inside the new enclosure of the group and release him after one or two days. This proved to be very effective in minimizing aggression triggered by a new male.

We established a trio of male squirrel monkeys at our institution. The animals lived in a generously spacious cage with numerous perches at different heights, platforms and hiding places. This group was a great success. We had one minor incident when the hierarchy seemed to change, but other than that these males were socially very compatible; they were always foraging, moving, and eating together as a coherent little group.

I will soon try to combine two female cyno triads into a single group of six; the animals are 4–6 years old. I have not worked with groups this size, so I am seeking any input regarding different introduction methods.

Each triad is housed in two quad cages connected with a tunnel; the plan is to house all six animals together in three quad cages connected by two tunnels. Currently the two triads are housed across the room from each other within visual contact.

Has anybody on the forum attempted similar introductions?

I wouldn't recommend trying to establish one group with these two trios; the animals are no longer young enough for that. It is my experience that combining two already established groups of adult females is very risky and can result in serious injuries of one or several animals, even if the two groups have been housed adjacently for several years.

Female cynos are such brats!

I agree; combining these two groups of adult females is probably not a good idea. If the animals would be 2-year-old youngsters there would be no problem.

Several years ago we established two groups, each of seven approximately 4-year-old female cynos. The animals were first familiarized in their future home pens by facing each other in transport boxes that were arranged in a circle. The pens measured approximately 12 x 8 feet [3.6 x 2.4 m] and were 14 feet [4.3 m] high. There were two wall-mounted perches at the back wall at two different heights, and two hanging structures. Aspen shavings served as bedding. The closed boxes were left in the pen for about an hour and then opened. The tech stayed in the room for several hours to monitor.

The formation of these groups was unproblematic and the seven animals of each group remained compatible for four years, at which point the groups were dismantled for reasons other than incompatibility.

I think the compatibility of these groups was partly due to the fact that the pens were designed in such a way that subordinate animals could not be cornered but always had escape routes from dominant counterparts.

We have created groups of up to five adult females with a lot of success.

Those of you who do a significant amount of social housing, how do you deal with pairing/grouping of sexually mature male macaques? What criteria, if any, do you use to justify not socially housing males if they have gotten into altercations with social partners? What level/frequency of confrontation do you use to determine that they are not or are no longer compatible, or do you continually attempt

PRIMATES 99

social-housing a male with different partners in hopes of finding a compatible match?

And one more question for the philosophers: is it better welfare to continue to attempt social housing with animals who get into fights that result in injury, or would it be better welfare to stop attempting to house certain animals with social partners, in order to avoid social stress and physical injury that can come along with social incompatibility?

Three documented strikes and you're out. We don't have the time and resources to try every possible combination.

We too have a "three strikes and you are out" policy for an aggressive animal who repeatedly injures his trial pairing partners. I do not feel that it would be fair to potentially injure more of our boys in attempts to find a partner for a male who really does damage to others.

We don't have a hard-and-fast rule for our cynos but follow a three-strikes line of thinking, meaning we attempt to pair a difficult male with three different partners. However, a male can be excluded from social housing if he attacked the other partner's body cavity—typically lacerations occur around the face, shoulders, upper arms, upper back, or legs—or if the wounds are life-threatening.

I have worked with quite a number of adult male rhesus pairs and found that partners were compatible during the first year after pair formation in about 80% of cases.

Pair incompatibility was triggered in some cases by husbandry-related factors or by other animals across the aisle, but most of the time it was related to variables that were not noticed by the attendant care personnel or by me.

Pair incompatibility became evident in four ways:

1. overt non-injurious, persistent aggression (ultimately leading to #3 if not noticed in time),
2. inadequate food sharing,
3. depression, or
4. overt injurious aggression.

I think it would be unethical to force two incompatible partners to live together in the same cage; therefore, partners were separated for good in scenarios 1, 2 and 3; they were then paired with other adult males or—and that was always successful—with naturally weaned infants from the breeding colony.

When the partners were engaged in injurious aggression—scenario 4—I first took care of the injury if needed, inserted a grated cage divider, and checked the pair's behavior very carefully for at least 24 hours; this requires patience and takes some undisturbed time. If the two partners showed clear signs of a dominance-subordinance relationship, I removed the grated partition; if their relationship was not clear, I separated the two for good.

It has been my experience that some males seem to get along with no other adult male when they are 4.5–6.5 years old; they are real rowdies and don't hesitate to provoke senior males—only to be beaten up. I think these guys—and especially partners who live with them—are better off alone until the rowdy reaches full maturity, or the rowdy can be paired with a naturally weaned surplus infant. It has been my experience that even the most belligerent rhesus male turns into a gentle, caring teddy in the presence of a little kid.

I am looking for information on pair-housing adult vasectomized males with intact female rhesus.

We have several such pairs, and they all work out really well. The vasectomy procedure is easy, and the pairing is a breeze.

I was able to keep several groups of adult female rhesus, each group with one vasectomized male. This housing arrangement was successful, except for the part where one vasectomy failed and we had a bunch of pregnant girls. Other than that we had no problems, and I think some of the groups are still together 5–6 years down the road.

It seems to me that previously single-caged male rhesus are easier to match up with a compatible companion than male cynos; is that true?

I have been trying to pair adult male cynos with each other for three days and all I am getting accomplished is perfecting my suturing techniques. My boys have not been successful at all. At least I'm getting an idea of who is not compatible. Potential pair partners knew each other and had audio/visual and limited tactile contact for about a month. They showed no signs of any aggression between them, even at produce times, so I finally figured the next step is full contact by removing the cage-dividing familiarization panel.

The two partners of the first pair immediately tried to kill each other. Needless to say they were separated; one of them required extensive repairs but he is doing well today. Now the two just glare at each other.

The other two pairs seemed to get along fairly well for several hours. It was obvious that they had established a rank relationship without fighting. Then, seemingly out of the blue, the monkey poo hit the fan. In the second pair one male just went bananas on the other for a couple of minutes. The injuries were minor and the two gave the impression that nothing really happened; there was a lot of grooming, cooing and hugging. So I left them together until the end of the day and then pulled out the one with bite injuries and stitched up a couple of spots. I left the two separated for the night and was hoping to put them back together next morning. Well it did not work out as planned; no fighting was going on, but the injured one was screeching and hiding every time he saw the other one, who acted like a bully and displayed unmistakable signs of aggressive intentions. I didn't take the risk but separated the two again, this time for good.

My third pair was doing well for several hours and then had a scuffle. The two settled down quickly after that dispute and seemed to get along with each other quite well; they groomed each other, followed each other and vocalized together. I was afraid to leave them overnight so I put the familiarization divider back before going home. First thing next morning, I removed the divider and all was just fine—at first. I stayed in the room for an hour, and then went into the adjoining room while keeping an ear for potential trouble. Again no problems for an hour. I left the facility for a short while, and during that time the two got into a fight. Well, that is an understatement; the little guy was fine but the big one had multiple, albeit not serious lacerations. I got him all cleaned up and don't really know what to do now. I'm just so tired of them getting injured but at the same time don't want to give up on having them live with a companion rather than be stuck alone in a single cage.

I wish I had more training and experience. Learning as you go along is fine for many things but not when ignorance and lack of experience can lead to the animals getting hurt. One thing of which I am not quite sure is how to determine correctly if two animals have established a stable, unequivocal dominance-subordinance relationship. Are there any red flags that warn you that two seemingly compatible partners are on the brink of a serious conflict? The three pairs that I am working with are sharing a room with females, and I am wondering if this situation could possibly be a reason for the males' intolerance of each other.

I have a feel-good-don't-give-up-trying story for you. We had an adult female rhesus who I would have put money on never to pair. She had a companion before, but when her mate got an implant she turned on her and the two had to be permanently separated. The female in question is/was aggressive toward all the other girls in the room, including the humans. She LOVES to challenge people. "She'll never be able to be paired," I said. She had bitten other animals before, challenged everyone and everything, and just would not get along with anyone. Then one day, her cage—old baboon cages where they could stick out their arms!—was pushed too close to another monk's cage. What happened was amazing. The two neighboring animals immediately hugged each other through the metal bars! Wow, the usually so feisty and aggressive female didn't want to bite the other monk who in turn seemed to actually like her!

Then we got our new cages, and we could pair these two females properly in a double cage. We paired them without any ado, and they have remained together as compatible companions for several years now. They're even going off to retirement together in a few short months!

I am sorry that these six boys are causing you such a headache! I have transferred approximately 100 single-caged adult rhesus males to quasi-permanent compatible same-sex pair-housing without encountering major obstacles. My experience may give you hints on how you could perhaps improve your success with these male cynos.

I always gave potential partners first the opportunity to get to know each other and sort out who is dominant and who is subordinate in a protected contact housing arrangement. I relied on strictly unidirectional grinning as a clear sign that the two have established a clear-cut dominance/subordinance relationship. Unidirectional yielding is another very good sign, but it is more subtle. Don't rely on dominance gestures (e.g., charging, threatening, glaring at the other partner), mounting or grooming.

On rare occasions, two males did not show clearly within the first 24 hours of familiarization that they had established a rank relation. When this happened, I did not proceed with pairing them but tested each one with a different potential companion.

When I paired adult rhesus males who had established their rank relationship during the protected-contact familiarization period, I always introduced them in the morning and when I saw clear signs of compatibility, allowed them to stay together also during the night. When you separate a new pair for the night, there is bound to be a risk of overt aggression when you introduce the two partners—again!—the following morning. It's important to allow new companions to settle without interruption into an amicable social relationship. Especially during the night, companions tend to huddle with each other, and that's what you want them to do.

Keeping male pairs—especially new pairs—in a room where they can see mature females is not a good idea. Most well-settled pairs can cope with such a challenge, but some pairs don't, and the consequences can be, as you probably have already witnessed, very traumatic.

If you come across adult males who just don't get along with anybody, stop trying and allow them to get a bit more mellow, even if this implies that they have to remain in single cages for another year or two. We cannot possibly force compatibility between two animals. Macaques WANT to live with another compatible partner or several partners, but the constraints of the lab can make it problematic to address this strong need in each and every case.

Thank you so much. I really struggled with the idea of leaving paired partners together during the first night; I was so afraid that I would come into a blood bath the next morning. I have lost two males in the past due to fight injuries and I hate to take chances, especially when I really do not have the experience with this. I think the first thing I will do is pull the females out, give the room a week or two to resettle and for the injured ones to heal, and only then try again. This time I will leave paired partners together during the night.

Our cages do not have grooming panels or protected contact cage dividers; I think that lack is a hindrance to the pairing process. I would like to get some fabricated, but I'm not sure what the best design is.

I have always worked with grated cage dividing panels. This allows the two potential cage mates to communicate with each other but prevents them from biting each other's finger tips. Visual privacy is not an issue during the familiarization phase, but once the pair has been established visual privacy can become very important for some pairs.

You may also consider introducing new companions not by simply removing the dividing panel, or taking the removable bars away, but by transferring them to a double cage in a *different* room in which everything is strange but the other partner, and no territorial feelings can interfere with the first direct meeting. Once the pair has turned out compatible, you can bring them back to their original, now interconnected home cages.

I have a few suggestions/tidbits/impressions based on my experience. It may or may not relate to your situation, but perhaps some of it will be useful.

I agree, get the females out! Test different male partners; start fresh without the gals around.

As a broad rule, I agree that it's best to avoid disrupting the developing relationship by separating a new pair at night. On the other hand, if you perceive specific contexts or triggers of fighting, partial or very brief separation needs to be an option. Generally, I don't think human micromanagement is very productive, but on some occasions it may be okay; this is my experience when we've noticed brewing social tension between two new companions in the course of the afternoon. Contact-grooming panels can be very helpful in such cases, because they make physical, possibly injurious interactions less likely, without disrupting visual, olfactory and auditory contact and communication between the two males.

One more thing, which may or may not be relevant to your situation: when we introduce two familiarized partners, we always make sure that they cannot be seen by another pair, especially one that is having a rocky time.

Using the same pair formation technique and the same housing arrangement, Lynch (1998) and Reinhardt (1994) tested 17 adult male cyno and 20 adult male rhesus pairs, respectively. Throughout a follow-up period of one year, partner compatibility was 94% for the cyno pairs and 80% for the rhesus pairs. These findings strongly suggest that previously single-caged adult cynomolgus macaques can be matched up with each other as compatible pairs as readily as adult male rhesus macaques.

Approximately what percentage of your facility's caged macaques are pair-housed? Please do not include animals of a breeding colony.

About 75% of our 65 primates [cynos and rhesus] are pair-housed. Social housing is the default, but of course there are exceptions—whether for research or medical reasons.

All our guys (41 male macaques) are singly housed. We're trying to get that changed, but it's a long, uphill battle.

Sadly, my reply is also 0%; our 39 rhesus macaques are all single-caged.

More than half of our 50 cynos are pair-housed. There are a few animals who seem to have problems living peacefully with another partner, but we don't give up and hope to get the remaining half of our colony also transferred to compatible pair-housing.

We have 157 monks: 47 cynos and 110 rhesus. All but one [98%] of the cynos are paired—and that's because we lost one paired animal due to a medical condition. About 50% of the rhesus are paired. I'm trying hard to come up with more compatible adult male-male rhesus pairs; it's quite a challenge!

I estimate that about 10–12% of our close to 1,000 rhesus macaques are pair-housed.

Currently, 98% of our 400 cynos are pair-housed. It is a constant work in progress, but we get a lot of support from the study directors to maintain our pairs. They will even keep the animals in pairs for group

assignment on study. We also request that our animals be already paired at the vendor; this really helps with our success.

Currently, 85% of our 440 cynos are pair-housed. The remaining 67 animals are exempt from social-housing for IACUC-approved research-related reasons or because an animal exhibits consistent social incompatibility with partners.

Research protocols sometimes require that compatible macaque pairs are physically separated for a limited time period (e.g., controlled food intake studies; sample collection from chair-restrained subjects; timed breeding). Obviously—and this is documented in the literature (Hennessy, 1997; Watson et al., 1998; McMillan et al., 2004)—partner separation is a stressful event that not only has animal welfare implications but can also skew subsequently collected research data.

If pair-housed macaques in your care have to be separated, (a) what do you do to minimize the stress for the animals or, if you don't have the authority, (b) what would you do to minimize the stress for the animals?

We use a wire-mesh partition or a solid Plexiglas panel. In either case, separation of paired partners is always for the shortest amount of time possible.

When pairs have to be separated for some time, we allow the partners to keep maximal visual contact (e.g., lexan solid panel) or maximal limited physical contact (e.g., perforated panels). I've worked with several pairs who have remained separated in this arrangement for up to two months and longer.

Reuniting them after completion of the project was always uneventful, probably because the partners had been able to maintain uninterrupted contact for the duration of the required separation.

Wire-mesh separators allow our pair-housed macaques to keep visual, olfactory and limited physical contact while one or both of them are assigned to studies that require physical separation of the partners.

Grated cage dividers are used at our facility; they make it possible for paired companions to stay in their home cages and keep contact with each other while they are physically separated during certain tests.

When we have to separate paired monks for feces/urine collection or for food consumption measurement over a period of one week, the partners are always allowed to stay in their familiar home cages and keep visual contact with each other through a wire-mesh cage partition. It is our experience that partners do not engage in aggressive interactions but do get along well with each other once we remove the partition after termination of the study.

We did have a pair of girls who did not properly adjust to being separated by the mesh partition while the study was going on. One of them would not eat well in this situation; it made me sad and showed me very clearly that being physically separated from the familiar cage companion, and not being able to groom each other during the day and huddle together during the night can be really hard for macaques.

Intermittent pair-housing of macaques

How many of you pair-house macaques intermittently? What is your experience?

Our cynos are separated with cage dividing panels every day shortly before the morning feeding; they are kept separated until after the afternoon feeding is complete. Partners are typically paired up for the evening by 2:00 p.m. We encounter no problems related to aggression with this arrangement, but it ensures that the two animals do not compete over food and get exactly the ration that is allocated for each one of them.

Our macaques are paired in large pens, but they also have access to home cages. We do have some animals who compete for food and others who will just not eat in the presence of their partner. In these cases we temporarily separate the companions in their home cages for the morning and again for the afternoon feeding, and pair them back again after they have had sufficient quiet time to eat their ration.

Once we form a compatible pair, we try to keep them together as much as possible. However, there are times when companions must be separated for medical or study-related reasons; this can be anywhere from a few hours to a couple days. To date, we have had no problems re-establishing pairs after a separation. Depending on the pair, sometimes we use a brief period of protected contact, but most of the time this isn't necessary.

If partners of an established pair have to be separated, it is our policy to re-unite them as soon as possible. If the period of separation has been a substantial amount of time (more than a week), we keep the partners—especially if they are four years or older—in a double cage with protected tactile access during their first night of being re-united. Once we see that the two have recognized each other as cage companions, we allow them to be together again uninterruptedly.

I had to deal on two occasions with compatible rhesus partners who started fighting with each other the moment they were re-united after a few days of separation. My conclusion was in both cases that the two partners didn't recognize each other quickly enough as buddies. As a result of these incidents, I made it a strict rule to give companions, who were separated for more than 24 hours, the opportunity to clearly recognize each other in a brief protected contact arrangement (lexan panels or grated cage dividers) before they are re-united. Since then, we have never encountered any aggression when partners were brought together after a few days or several weeks of separation.

We have paired rhesus macaques, who are separated, with grooming-contact bars—some overnight, others during the day—for research-related reasons. When they are separated, the companions keep visual, olfactory and partial physical contact with each other. We have had no issues with re-uniting them the following day or after several days of separation. The only problem I've had was when companions had no visual access for an extended period of time (more than two weeks) and were then re-united without any preliminaries. I think such cases need to be treated like new pair formations in which partners are first carefully familiarized in a protected contact environment before they are released into the same home cage.

Pair-housing macaques of different species

Occasionally, single-caged rhesus macaques are transferred to pair-housing conditions with another macaque of a different species. If you have dealt with such pairs, were the rhesus partners dominant or subordinate in cases in which both partners were of the same age group?

Since we have a limited pool of potential pairing partners at our small university, we have done this a few times with rhesus and cynos (both male and female pairs). In every case, the rhesus assumed the dominant role. We published a short report on our experience (DiVincenti et al., 2012).

We have a 14-year-old cyno female who lives with a 5-year-old male rhesus as a compatible pair.

This pairing turned out so well! The female is a strong, confident bully and had been hard to pair because of her unreasonable aggression toward other macaques; she weighs 22 pounds and is rather large-boned for a cyno. This young rhesus explained to her that he was the king but that he would allow her to do his laundry. She asked him how he wanted his socks folded (grin).

The two have become really great companions.

environment, and that the two are grooming each other. Perhaps more surprising is the observation (shown above) of a baboon affectionately grooming a vervet in an African bush setting.

Oral dosing of monkeys

Has anyone experience with nasogastric or orogastric intubation of capuchin monks? One of our investigators is looking for a reliable oral dose administration route for these animals. Apparently, when the investigator mixed the compound with the food, the test results showed a conspicuous discrepancy with the literature. For this reason, we are looking for an alternative, perhaps more accurate PO [per os] dosing technique.

We have one elderly female pair of a rhesus and a stump-tailed macaque. From what we can tell, the rhesus is dominant. They've been a compatible pair for several years now!

It is not very surprising that a cynomolgus macaque can match up with a rhesus macaque as a compatible pair in the captive

We perform nasogastric intubation for oral dosing of our cynos very often. Our animals are already sitting in restraint chairs prior to and also during the procedure. We first acclimate them to gentle-and-firm manual head restraint and smooth nasogastric intubation until they no longer show any signs of discomfort or stress. When you work with an acclimated animal, restraining and dosing him or her takes no more than one minute. I am not aware that any of our animals ever got harmed or injured during this swift procedure. Two people are involved in it, the doser and the restrainer. I have never performed the dosing but have seen it done many times as I am usually the restrainer.

We use the chair's arm restraints to keep the animals from grabbing at the doser and the restrainer. It is my job to hold the monkey's head in a steady position that allows the doser to intubate the animal smoothly and administer the drug. Our animals are so well habituated that I do not really have to restrain them while I am holding their heads. They have learned that they can trust me and that I expect them to look up and stay still; and they do stay still with me!

There are only two of us who are dosing our cynos. We are also responsible for chair-training and target-training the animals; we are also feeding them and provide the foraging and structural environmental enrichment. We are working with our animals every day and have developed affectionate, mutual trust relationships with all of them. This, probably, is the very condition that makes it possible for us to dose them without causing stress and without injuring or harming them in any way.

We have been very successful at dosing conscious, albeit chair-restrained rhesus monkeys via orogastric intubation. We are making use of an in-house-made bite bar to keep the animal's mouth open and prevent the animal from blocking the tube with her or his tongue and from biting the tube. We use 8-French tubes; their relatively large diameter makes it certain that the tubes cannot be mistakenly pushed into the trachea. Most of our monks don't seem to realize they were actually dosed; it's so quick.

This is very interesting to me. I have tried to dose monks orally but haven't had much success because they bite the tube and/or will push the tube back out of their mouths with their tongues. I did consider using a bite bar, but feared they might chew so hard on it that they crack a tooth or suffer some other oral injury. Your report is giving me hope; I will try to follow your technique.

I remember reading somewhere about tricks for getting nasty tasting oral medications into fussy primates. We have a few rhesus who are so finicky—and quickly figure out what we are up to. Fresh produce tends to be the best bet (e.g., strawberries, bananas), but for bitter tasting meds it doesn't always do the trick. We have tried peanut butter, Twinkies, Fruit Roll-Ups, baby food, applesauce, etc., but there are always a few monkeys who keep you on your toes. Any suggestions would be greatly appreciated!

If caloric issues aren't a concern, Kool-Aid powder is an excellent vehicle to disguise very bitter liquids or crushable meds. I had luck using orange flavor in squirrel monkeys and marmosets. They'd drink it right from a syringe, or I could mix it into their water bottle.

When our rhesus or cynos come in, the husbandry staff and/or the vet techs visit them regularly in order to establish a friendly relationship with each one of them. During these visits yummy things such as Kool-Aid, juice, honey or applesauce are offered to the animals from syringes. Most of our current residents can and will accept oral meds via syringe if needed.

Your approach of informally training monkeys to cooperate during oral drug administration is great! It is not complicated or difficult, yet it can make life both for the animals and the humans so much easier, while fostering at the same time human-animal relationships that are based on affection rather than fear.

As part of my enrichment program, I started out teaching my rhesus macaques (males and females) to take Margarita Mix from syringes. I squirted the liquid at the monks to hit their hand and after they licked it off, they inevitably came to taste a drip from the syringe. It took two sessions of a couple minutes each to teach the guys that the syringe contained *good stuff*. I used 20 cc syringes because it was easy for the monks to pull smaller syringes with their teeth into the cage, and some monks got frightened when I approached them with small syringes, probably because these were commonly used to give them injections.

The training of the monks paid off when the researcher wanted to oral-dose them and I informed him that it could be done without much ado and without the need to anaesthetize the animals.

We have really good luck with a spoon. We mix meds in various food items that best mask or complement the taste of the drug; strawberry yogurt works for most commonly used drugs. If a drug is very bitter, we mix it in something else that also has a bitter flavor. Coffee does the trick in such cases; yes, ALL our macaques like coffee. Coffee with a tad bit of Coffee Mate fat-free vanilla creamer effectively masks bitter drugs, and the animals lick it from the spoon without much hesitation.

We first train the animals to lick the drug-masking liquids from a spoon without using their fingers; otherwise they may grab it, smear it on their body or my arm, or spill it altogether. It doesn't take much time to teach them how to use the spoon as intended. We then mix the med in the liquid of choice and offer it again with the spoon.

[This simple trick with the spoon may be applicable not only for primates but also for rats. When searching in Flickr's Creative

Commons for photos of rats I happened to come across this picture:

The photo with its caption "Balboa gets his meds" suggests that, if mixed with the right decoy, some drugs can be administered to rats also in a stress-free way with a spoon rather than with a distressing stomach tube.]

Whoever would have thought of giving monkeys coffee? Have you experimented with regular and decaffeinated coffee?

We use such a small amount that it really isn't a concern; but FYI [for your information], we use regular coffee as that is what we have available around here.

I am wondering if coffee would work with the notoriously difficult-to-mask Flagyl.

Flagyl (metronidazole) is the reason why I started playing around with different compounding options. You see, in my earlier years I was a compounding pharmacy tech and learned about *what* goes with *what*. You wouldn't get a pleasant taste if you mixed sardines and ice cream, or vinegar and milk. Does that make sense? YUCK! The Flagyl is bitter and lingers in a bad way, so by adding something also bitter but in a good way, the two tastes blend; and then adding a hint of sweet that would also blend, making the cocktail palatable. I use 2/3 coffee with 1/3 creamer. In some cases I even add a drop of vanilla extract to that mix. Now SOME monkeys still will not take it, but most do. For the ones who won't take it in the coffee, I mix the Flagyl with ketchup; a few accept this cocktail.

I need to mention: ALWAYS offer the coffee or whatever you are going to mix the meds with several times pure, before you add the meds! The animals will quickly develop a taste/craving for it, especially for the coffee, and will then put up with the little *extra taste* of the meds, just to get what they really like.

Coffee, that's something I never would have guessed monkeys would like. We'll certainly need to try that one! Thanks for sharing.

I was wondering how those of you who do what I would call cooperative oral dosing—versus dosing under forced conditions—are able to ensure that the subject ingests the correct amount.

It was always my wish to use a more cooperative method, as I had never had any of my monks refuse treats or juice. The PIs were strictly against it, arguing that I could not guarantee that the total amount of hidden test drug was actually ingested by the monkey.

I've also run up against this many times.

When I oral-dose one of our monkeys, I know for certain that the animal has ingested all of the compound when she or he has licked the spoon clean and swallowed one more time.

We rarely encounter a monkey who won't lick the spoon clean. Once in a blue moon a monkey will not lick everything from the spoon. In this case we mix whatever is left over with one or several different favored food items until the monkey has taken the entire dose. This may take some time.

I would suggest that animals in laboratories should be given the chance to cooperate with the licking of the test concoction from the spoon—or from the syringe tip. Only if they fail to lick the spoon clean or empty the contents of the syringe should gavage with a gastric tube be considered. Why not allow those who WILL lick from a spoon or from a syringe do so? Why would anybody use a nasogastric tube in preference to letting a monkey lick from a spoon or syringe tip, when nasogastric intubation is stressful for the handling person and distressing for the animal?

I trained 43 marmosets to drink their doses via syringes with blunt tips. First, I tested many flavors and came up with the following favorites for our colony: Splenda in 5% or 10% solutions, maple syrup, blueberry syrup and raspberry syrup; the latter two syrups are also available sugar-free.

We were able to obtain the cooperation of all 43 marmosets and dose them during studies—most of the animals right through the bars without missing a single drop! We dosed the colony this way for four years, both during acute studies and chronic studies extended over a time period of up to 14 days [Donnelly et al., 2007]. The rationale behind the training was twofold:

1. No one in my group had ever gavaged a marmoset, so we were all worried about the risks of damaging the esophagus of the animals.
2. It was important to the PI that the marmosets were not unduly stressed during the oral dosing.

We are now trying to train rhesus macaques to fully cooperate during oral dosing. Rhesus seem to be much smarter than marmosets, so it is quite a challenge to trick them. At first we tried to dose them with various flavors. They were great and licked the tasty liquids without fail from a syringe tip—until you

added a compound; that was the end of this avenue. So far, hiding certain drugs in yogurt has worked well. We give them the yogurt-drug-mix in paper cups, leave those for about 15 minutes and then check; by then, usually all is gone and the cup has turned into a little, chewed-up paper ball or is simply licked clean. If the monkeys spill some of the mixture, they usually clean it all up with their hands and then lick their hands at great length. We are also in the process of trying pudding with frosting and applesauce in little tiny ice cream cones so that the animals can eat the whole thing.

I love this discussion but it gets me so frustrated to read that others are implementing these better methods, which are far kinder to the animal, without PI issues, when I would be faced with such unreasonable road blocks. As I'm sure others have encountered, you spend months working with these animals, gaining their trust and cooperation, just to have a group of techs fly in one day, shove a tube down their throats and then leave. Needless to say, your animals are back to being terrified of any human they see, they are angry and perhaps even aggressive.

I followed with interest the discussion on oral dosing. We have to continuously administer substances to our vervet monkeys but rarely need to actually dose orally due to the way we feed our monkeys. I know it would not be feasible for most other facilities because they are feeding commercially prepared pelleted food (chow), but for what it is worth, our method has been described in a journal (Seier et al., 2008). We also compared blood concentrations after gavaging and using the above method and found the latter acceptable even for PK [pharmacokinetic] studies, for which we use it now extensively.

The food [facility-made maize meal-based diet] that the vervets receive in the morning weighs 100 grams [3.5 oz] and for the PK we take off 30 grams [1.1 oz] and blend the compound to be administered into that. A small amount of honey can be added where necessary to mask flavors. The portion can take dry and wet material. The monkeys are hungry in the morning and eat everything immediately without storing anything in cheek pouches. I should say that our vervets—unlike macaques—are not too prone to store food in cheek pouches at the best of times. For other studies we blend the compound into the entire food ration. Wastage is zero in the 30 gram portion and less than 10% in the 100 gram portion. The latter is loss during handling rather than active wasting. During studies we weigh the consumption of every animal to establish compliance and can even adjust the dose for that.

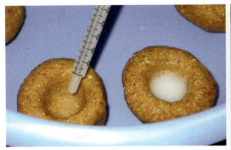

The photo above shows how a 30 g portion is filled with a liquid compound for a PK study. After filling, the portion is pinched, closed and gently kneaded to produce a homogenous consistency. [The vervets eat these baited doughnuts without hesitation.]

To optimize the animals' handling and minimize the drawback of the oral gavage, we developed a refinement for conscious cynomolgus macaques. After implanting a subcutaneous port, a surgically-placed gastrostomy (SPG) was completed to afford access to the gastric lumen and enable the administration of substances. The device was left in place for 2–12 months in 11 macaques. In five cases, the SPG was used successfully for 8–12 months, until the experimental endpoint was reached. In six cases, the SPG had to be removed earlier due to local infection at the implant site (Fante et al., 2012).

What you describe seems to be a very smart Refinement.
When you administer the drug via SPG twice a day over a period of several weeks:
1. *Can the monkey stay in her/his familiar home cage?*
2. *Is the monkey somehow restrained while you administer the drug?*
3. *Do you need a second person to help you do this frequent procedure?*
4. *Is the monkey sharing the cage with another companion?*
5. *I assume you tried administering the drug via syringe or other methods of direct oral feeding; were all your attempts unsuccessful, hence you resorted to the alternative SPG technique?*

1. Yes, the monkey stays in her/his familiar home cage.
2. Our approach requires only gentle pressure with the squeeze-back mechanism to bring the animal to the front of the cage in order to access the injection port. With time, almost all monkeys become used to this procedure, sometimes coming spontaneously to the front of the cage and waiting for the treatment to begin.
3. Yes, a second person is needed during this procedure.

4. Yes, the procedure is compatible with the presence of another cage companion.

5. We used oral gavage over a period of several years and we still use this method for short-term experiments without any complication. Nevertheless, orogastric intubation is inherently stressful for monkeys; in our experience, the animals are unable to adapt to this procedure. In this light, we were searching for an alternative refinement technique.

Your Refinement technique is ingenious. I am particularly impressed that this alternative approach of oral drug administration allows the monkey to remain in her or his familiar home cage, and that you are keeping these animals in a compatible pair-housing arrangement.

Preparing monkeys for handling procedures

What are the options for preparing monkeys for training protocols that will teach them to cooperate during handling and husbandry procedures?

I have managed to create some time for target-training the monks in my charge. I began carrying the target around the facility in every monkey room during routine rounds for about one week. This second week I began bringing the target very close to each cage while asking the monks to touch it. I had 22 animals yesterday who figured it out! It's great to watch them as they pick up on the game, hear a click as soon as they touch and then promptly receive a piece of veggie or fruit.

Today the number increased to 39; some of the animals are outstanding and already touch the target every single time I ask for it. It's amazing to see them figuring out what you are asking of them, and once the food rewards start coming, they really catch on! Some of the more timid monks are watching their cage mates *play* with me and before long are observing intensely how their companions are getting all these lovely grapes, pieces of apple, pieces of cucumber and other goodies. It's a lot of fun!

I find myself spending at least two hours every morning playing this game with all of my animals. By doing so I am building a close bond of trust with them which, ultimately, is the foundation of any successful training program.

All it takes is to always carry a clicker and some treats with me. Once the monks have made the association between the click and the treat/reward, I can click whenever I see a behavior or posture that I want to reinforce for a specific training goal, for example unintentionally presenting a thigh— eventually for injection.

To prepare the animals for training does not require a lot of time but it offers extremely valuable *enrichment* both for the animals and for the attending staff, while at the same time fostering a trust relationship between animal and handler, the basic foundation of any successful training project. Once an animal is no longer fearful of humans, has been prepared to kind of work with the target, or to associate the click with a favored treat, the training itself becomes so less time-consuming and much easier and more fun.

Such preparatory training steps can be integrated into the daily animal-checking routine without undue extra time investment.

It is my experience that rhesus macaques can learn many things once the bond of trust is there. The click and subsequent treat during routine rounds and clinical observations is a great trick; it's quick and easy and you will literally see the animals change before your eyes. I now have monks who exhibit all types of behaviors, including presenting the hind quarters for treat rewards. Some present for being petted; I could stay there all day grooming various parts; they enjoy it so much!

Based on these routine informal interactions with my monks, I have trained many of them for pole-and-collar and subsequent chairing, blood collection, walking on the scale for weighing, and entering and exiting a play cage.

The key for any successful training is mutual trust. You can shape various behaviors later on, once you have established that trust relationship with the monks; trust is so important. Trust also makes it safer for personnel handling the animals. A monkey who trusts you has no reason to be afraid and try to scratch or bite you in self-defense.

I very much agree with you when you emphasize that the development of a mutual trust relationship is THE key to success when conditioning, training or simply working with non-human primates or any other species held in captivity. Yes, you do need some extra time to develop such a relationship, but it pays off in easier and faster training and, ultimately, in scientifically more reliable data

that are collected from animals who are not experiencing intense fear during the data collecting procedure.

Monkeys cooperating during procedures without formal training

Who has worked or is working with non-human primates who have learned, without formal training, to cooperate during a handling procedure?

I have one recent experience that I can share.

I am in the process of forming iso-sexual pairs from a group of nine adult male cynos. Thus far, I have established two pairs. One male was bitten by his partner and has extensive damage on his left hand. I am now waiting to get vet clearance so that I can try pairing him with someone else. In the meantime I am monitoring the male's hand very carefully.

The other day we were sitting close to each other and I was talking quietly to him while displaying my left hand in a way he would have to do with his injured hand so I could examine it. To my amazement, he copied me! He raised his left hand and I was able to inspect the wound. I had never asked him to do that before; he simply moved his left hand in the same manner I had moved my left hand. I then tried with simple gestures to communicate to the male that I would like him now to present the left arm through the feeder-box opening of the front panel of the cage. He seemed to immediately understand what I meant, and put his arm through the opening. I gently held his hand and looked it over and then rewarded him by grooming his wrist and lower arm and offering a treat.

Yesterday, I visited these nine males. When I approached the male with the injured hand, he started putting his arm into the feeder box!

It was an amazing experience!

As a veterinarian I was privileged to deal with monkeys, cats, cattle, deer and birds who spontaneously held still while I treated a wound, administered an ointment, removed ticks, cleaned an infected eye, or removed an irritating foreign object. It seems to me that these animals somehow knew that they can trust me 100% and that I wanted to help them, so they relaxed and allowed me to do what had to be done.

It is my experience that some technicians/caregivers develop a trust relationship with individual monkeys from whom they draw blood samples relatively often in the traditional manner, i.e., using mechanical restraint either in a squeeze-cage/box or in the animal's home cage with the squeeze-back. For the individual animals, the blood collection procedure gradually develops into a predictable, hence no longer anxiety-inducing procedure, and they finally start cooperating without any formal training. You could say that the care personnel provide the proper ambiance for the monkeys to learn by themselves to cooperate with the handling person rather than resist.

I know of one adult female rhesus (Star of Cowley) who, without any formal training, has learned to come forward when her handler, Doug Cowley, opens the cage door slightly, and then position herself in such a way that Doug

can take a femoral blood sample without any ado. Certainly, Doug rewards this female after each blood collection with some raisins. I also remember two adult male and one adult female rhesus and one adult female stump who, without being mechanically restrained in the squeeze cage on the hallway, stuck out one of their hind legs to allow Doug or another handler, Russell Vertein, to draw blood from the saphenous vein. Again these animals received some raisins upon returning to their home cages.

These examples show how very important the animal handler's role is. Unfortunately, this is often not appreciated, so no provisions are made to ensure that these people have some time to just be with the animals in their charge, and that they can take samples from them without being time-pressured.

To be realistic, I do have to add here that I have also worked with animal caregivers who were callous and should never have been hired to work with primates; the animals predictably freaked out whenever these individuals approached them. To formally train such fear-conditioned animals takes

PRIMATES   117

a lot of patience and sensitivity in order to gradually gain their trust and then start with the actual training.

Monkey see, monkey do

Has anyone had experience with monkeys learning or picking up a certain behavior by seeing another monkey or a video of a monkey performing that behavior? How feasible would this be as a training tool in the laboratory?

I chair-trained two pair-housed male rhesus macaques and started with one partner while leaving the other one in the same room so that he could watch and provide social support. It seems to me that the onlooking partner partially learned the training steps: when it was his turn to be chair-trained, he knew immediately how to sit in the chair, and he also accepted treats as training reward right from the beginning. Watching the training procedure may have encouraged him to imitate his companion's responses, which made him cooperate more readily.

Here is another example of imitation: Two adult male rhesus were housed next to each other without visual contact, but they could see other conspecifics elsewhere in the room. Male A would routinely present his butt for grooming to most personnel that entered the room and was met with lots of attention and treats. Male B was trained to offer different body parts for grooming, but he didn't present his butt and would act aggressively if personnel tried to groom a body part that he did not offer.

Cages in the room had to be moved, and it so happened that these two males were now neighbors facing each other across the aisle. No more than a week after this new cage arrangement, male B started to present his butt exactly in the same way as male A did, and after receiving the same attention by the personnel he added this behavior to his repertoire; it became his favorite trick. He was no longer aggressive but presented in order to get his reward.

That these guys do learn this quickly from watching another monkey's behaviors makes it very likely that a well-done video could serve as an effective learning tool for them.

If you agree that chimpanzees are not necessarily more adept than macaques to learn via imitation, you will probably be interested in this abstract by Lambeth *et al.* (2000): "Subjects were 10 adult chimpanzees living in two groups. Five females were exposed to a 10-minute videotape of female chimpanzees being positively reinforced for successfully urinating into a cup. Immediately following videotape exposure, these subjects participated in a training session." On average, experimental and control subjects received 56 minutes of training. "Subjects with videotape exposure successfully responded to the command to urinate in significantly less time than did controls. ... Four of the five experimental subjects urinated into the cup in a mean of 5.75 minutes, while the fifth subject never urinated during the training sessions. Only two of the five control subjects urinated into the cup during training sessions (mean time = 43.32 minutes)."

I am sure rhesus, cynos or stumps can also learn through direct or indirect imitation, via video presentation; to my knowledge this has yet to be documented in the literature.

The *Cebus* we use at Helping Hands Monkey Helpers for the Disabled learn by imitating what the human trainer does: monkey see, monkey do. This ranges from simple tasks such as touching a lever to rather complex tasks such as inserting a CD into a computer and then pushing *play*. Not sure if video images would work, but if a monkey was interested in the screen's content, why not?

I have had monks learn not-so-nice behaviors, such as poop smearing on cage walls, from others.

This reminds me of a recent observation I made in our sanctuary.

We acquired a 10-year-old male rhesus with serious mental issues. He's a neurotic pacer and, according to the previous owner, has been this way for about 10 years. We put him in an enclosure across from our 4-year-old male rhesus who is extremely smart. I thought that perhaps the new "mental patient" might learn some "normal" behaviors.

The healthy young male watched attentively as the new roommate paced. By the second day he was also pacing, touching the back wall in precisely the same stereotypical way as the "mental patient"; he was just a bit clumsy about it. I must admit, it was funny to watch this smart guy acting at being neurotic. Needless to say, I rearranged the cages to make sure that he could no longer be seduced into pacing; he did stop!

I had two adult male cynos occupying pens across from one another. One was very curious and got into everything. He quickly learned how to let himself in and out of his home cage that was attached to the pen. He would open the door, slide under, and then close it behind him. The door was usually propped open, but he seemed to prefer having control of the door, so we let him have his way. His buddy across the room was more of the couch potato type; he loved to watch what the other guy was up to as long as it didn't require him to move.

One day I was wandering by and saw the "lazy" cyno messing with his door. After a couple of days the animal care staff reported that he had learned the same behavior as his roommate and started letting himself in and out of his home cage.

After that, we started to notice the lazy guy picking up more of the curious guy's tricks—for example, the quickest way to dismantle a peanut butter jar. Lazy guy would spend quite a bit of time chewing on the bottom and trying to tear the plastic apart, while curious guy swiftly pried the lid off with his canines and got to the peanut butter. That smart trick got imitated by lazy guy pretty quickly, probably owning to the fact that a favorite reinforcer was involved.

Our cyno, Mandy, has come up with a sign language of sorts that indicates she wants to be groomed by attending personnel. She extends her hands toward the subject she wants to groom her, makes full eye contact and quickly and repeatedly moves her little fingers up and down in typical grooming motion. She usually gets what she wants, which means her gesture is reinforced. The cyno across from her has been observing this for nearly two years, and just a few months ago has started copying Mandy's gesture *not* toward the humans, but toward Mandy. Now the girls sit across the aisle from one another and do this back and forth. I guess, it is entertaining for them and it sure is amusing to watch.

I have a male rhesus—one of my favorite guys—who does similar things: he lies on the floor and looks at his belly, scratches his belly, intently turns his head and looks at you, looks again down at his belly, looks at you—like *what are you waiting for?*— then looks down again at his belly, until you finally get it and start grooming his belly. His caretaker conditioned him to this by addressing him with *ruble belly*, a sort of baby talk. This verbal cue does the trick for anybody who approaches his cage.

But I also have cynos who imitate behaviors that one of them initially learned from a person: one of the care staff conditioned one of his favorite monks to respond to *nice-nice* by gently touching, stroking and grabbing the person's finger, then making eye contact; the person then rewarded her little friend with a food treat. Over the course of time, other monkeys in the room copied this particular monkey's behavior to receive the attention of their human caretaker.

Last night I tried it and all but one did it with me, even though it was the first time I asked for *nice-nice*.

I've had many rhesus through the years learn various things from others that would get them human attention, a treat or a toy. It creates an amazing human-animal bond of trust, which helps when training them to cooperate during procedures. I do believe they learn new tasks faster from the people they trust. Trust is incredibly important!

I'm in the process of training two paired rhesus right now to pole-and-collar and chair, and I ask them to station themselves at the front door at the beginning of each session. A few days ago, the submissive partner started doing an adorable head bob-bow during this session; we actually reinforced this behavior with a treat because it is helpful when poling the animal. Today the dominant partner did the same bob-bowing gesture at the cage door, hoping to get our attention—and a treat. It was quite cute; he obviously had learned that behavior by imitating his submissive partner's rather dainty gesture.

Training monkeys to enter into a transfer box/cage

Does anyone have experience with training marmosets to enter into transport boxes? If so, how long did it take you to see consistent results?

The first step will be the most time-consuming, that is to gain the marmosets' trust. Food rewards will definitely quicken the process. Macaques warm up to people with food treats quickly; the same may be true for marmosets.

I've been reading up on training marmosets, and I'm getting the impression that food will win them over, but I fear that by the time they accept taking food from us and are comfortable, it will almost be time for them to go; these animals will be on a 3-month study.

Once you find that *winner* food treat, you can reserve it only for bonding, and of course later for boxing. What helps us with macaques is keeping the same person offering the rewards and performing the initial training/acclimation.

Go with marshmallow—mini marshmallows will win the heart of any marmoset.

Yes, marmosets LOVE marshmallows; this circumstance makes these treats excellent positive reinforcement training rewards!

Years ago we trained our marmosets to enter a transport box. They were mixed age groups (mom and 1–2 daughters). They readily came up for treats; so that was a very good start. With daily 5- to 10-minute-sessions, in less than two weeks we were able train them to reliably enter a transport box. We baited the box with mini-marshmallows, which are a real crack for marms. It was easier to get the whole little group to enter the box than individual animals.

I have trained marmosets for hand-catching and per os dosing. I worked with younger ones and had time to form bonds with them; older marmosets, if never handled, can be quite a challenge. Marmosets are quite timid and get easily afraid when confronted with new objects such as gloves, boxes, toys, etc. I used to place new items on a cart in the middle of the room for a week or so, just to give the animals a chance to get used to seeing these things.

For me, hand-catching the marmosets—followed by a food reward such as maple syrup sucked from a syringe—was easier and much quicker than training them for transport boxes.

I will have a month to work with the animals before the research starts. I conclude from your comments that hand-taming, hand-caching and rewarding with favored treats may be the way to go.

Spend lots of time with them and you will see their little personalities come out. I had quite a few little hams; they used to sing like birds when I brought in live meal worm containers—they loved them!!

You do have to be careful when catching the animals, not to twist or pull too hard; marmosets are very fragile! I gently but firmly grab them around their waist and pull them very carefully until they give up holding onto the cage and relax in my hand, or I slide my free hand under their chest and they will typically let go of the cage and relax in my hand.

I must add that I am also getting slight resistance from the husbandry staff, since this is so new and many people doubt the ability to train the marmosets; hopefully I'll be able to prove them wrong!

We trained our husbandry staff on how to hand-catch the animals for cage changes, and they did great!

How do you train macaques to enter into the transfer box without making use of the squeeze-back, for example, in standard cages without squeeze-backs, or in pens?

I am working at a biomedical research facility; this issue brings up a lot of frustration for me here. Almost all of our macaques are single-housed, and for the most part they are not trained for any cooperative behaviors. I do not know why the research staff does not take the time to train the animals before they work with them. It breaks my heart to see monkeys squeezed up and threatened with brooms in order to make them enter transfer boxes, because no time is allocated for training the animals to cooperate rather than resist during common husbandry procedures, let alone research-related procedures.

There should be no need to threaten the monkeys with brooms! The animals will have a very difficult time trusting anyone when

the research staff treats them in this manner; monkeys are very smart but also sensitive. Only when humans have earned their trust will they be willing to work with them, rather than being filled with fear and apprehension when they see a person approaching them. I understand why your heart is broken, mine is, just reading your observations.

With the use of a banana or other fruit placed into the clean cage, our rhesus macaques learn quickly to exit their dirty cage and transfer via a tunnel into the clean cage. This works very well and most of the time the squeeze-back is not needed, except for a few exceptional animals. With some of the shy young guys it requires about three cage changes—which are done every other week—to also get them to jump reliably.

We put a few PRIMA-Treats or peanuts into the weigh box, which is temporarily attached to the cage front, to entice our young, inexperienced rhesus macaques into leaving their cage and entering the box. Most of our adults have learned the trick and enter the box spontaneously, knowing that they will receive their reward upon re-entering their home cage after their weight has been taken. I find adults are very nice to work with; they tend to learn faster than the youngsters.

Our colony of cynos is fairly small (about 27 animals total) and all animals are singly housed. They are moved in little groups into a test enclosure for behavioral monitoring. Transport boxes are hooked to the front of their cages; they exit into the box and are taken via cart to the test room where they are released into a special enclosure for four to seven hours, depending upon the testing day. At the end of the monitoring phase, the transport boxes are hooked to a gate on the enclosure and the animals run into the boxes according to their social status within the test group. Upon returning to their home cages they are each given a slice of fruit.

No formal training was done other than getting the monkeys used to the transport box routine; this took only a few trials, once daily for two to three days in a row. These animals are very smart, and discover quickly that a reward is waiting for them in the transfer box; they are really good about going into the box. If any one of them is kind of stalling, I offer a special treat incentive such as sunflower seeds in the transfer box. I once had one cheeky cyno who stretched into the box from his home cage and retrieved all the seeds and then positioned himself back in the cage, eating the seeds right in front of the open box. He was a trickster, but when I held up an apple slice for him to see, he immediately dashed into the box and let me lower the guillotine door.

It's been way more than a decade since I worked with cynos but when I did, it hardly ever happened that we had to make use of squeeze-backs. For cage changes we had a tunnel that connected the dirty cage with the clean cage; placing seeds or fruit into the clean cage was usually sufficient to have the monks move quickly into the new cage. Some were hesitant, but just standing back was usually enough to make them move through the tunnel.

For getting our cynos to enter transfer boxes, I used a lift-stand to hold the box. Those animals who promptly moved into the box and let me close the door were rewarded with an apple slice. For those who were hesitant to enter, I just moved a bit away and waited for them to enter the box. For some, I had to wait 10 to 15 minutes but usually it just took a minute or two until they decided to move. If these animals stayed in the box and let me close the door, they too got their fruit as soon as they had returned in their home cage. For those animals who didn't stay long enough in the box but jumped back into their cage before the door could be closed, I started the session again by moving a few steps away and letting them re-enter and stay in the box until I had closed the door. Usually this exercise had to be repeated about five times before they would stay in the transfer box long enough for me to close the door. I would then release them back into their home cage and give them their reward.

The transfer box training was rehearsed every other day for one or two weeks, depending on the individual monk's performance.

For the next step, I rolled the animals in the transfer box on a cart into their observation room or into their work chair where they were given, again, an apple slice. I waited with them for about 20 minutes; during that time, nothing was done with the animals; they could simply relax. After that, I boxed them up again and brought them back to their home cage, where sunflower seeds were waiting for them as a reward for cooperating with me. We repeated this five times before the animals were signed off to begin their tests. I didn't have a single monk who resisted the box transfer after the first test run.

The key for success, especially with older

macaques, is to give the animals sufficient time to recognize that the transfer box is not a trap. So you can't be in a hurry to get them to move into the tunnel or jump into the box. If you're stressed, chances are the animals are also going to be stressed and both parties will have a hard time with the training. It's always a good idea to not even start a training session under such conditions.

This simple training protocol that you described can be applied by anybody who wants to get macaques to enter into transfer boxes without much ado.

I assume you always worked with the animals alone, i.e., nobody else in the room and no commotion out in the hallways.

I definitely worked alone to avoid getting the animals frightened by too many humans in their room. The monkeys were used to daily procedures being done by a single person. Anything out of the norm is likely to make the animals suspicious, perhaps even alarmed.

As for commotion in the hallway, in a large facility there's almost always something going on—people passing by, carts moving, etc.—but as long as this is not startling for you, chances are, it's not startling for the monks either. If something unexpected does happen, you just sit/stand back and wait for all the animals in the room to settle down again and relax; only then is it reasonable to proceed with the training.

Is anybody on the forum in a position to share experiences with training squirrel monkeys to cooperate when you want them to leave their cage and enter a transfer box?

In my experience, adult squirrel monkeys are just as easy to train as macaques to move through connecting tunnels during cage change, but juvies tend to be devious. They'd run right to the edge of the tunnel, then sit there grooming themselves or picking at seeds, or sitting right in the middle of the tunnel rather than going all the way through to the new cage. Having a favorite piece of fruit that you place so that they have to go all the way through can help, but if you're not quick, they can grab the fruit, turn around and run back into the tunnel to eat it. I've not worked with them in years, but I still laugh thinking about the little games they'd play to make me wait until they were finally ready and decided to walk into the new cage and stay there while relishing their food reward.

We have trained squirrel monkeys to jump cages and have started training them to enter a transfer box for cage change.

Squirrel monkeys can be trained; it just takes a lot of patience because of their natural inquisitiveness and high energy levels. I really enjoy working with them but man, can they be frustrating, just like trying to train a toddler! They know what is expected of them but nothing will make them move any faster.

Years ago I volunteered at a local zoo and worked with the squirrel monkeys. We used meal worms as rewards for shifting into their indoor enclosures or for entering into transfer boxes. They loved their rewards so much that they probably would have gone anywhere for them; they grossed me out completely! I think with squirrel monks, as with all other primate species, finding the right food motivation is the key for success!

Training macaques to cooperate during blood collection

I have three cynos who have multiple weekly blood draws that are currently being done when the animals are sedated. This schedule will continue for at least one year. I would love to train these animals to present their legs and was hoping maybe someone on the forum can share practical advice. I am familiar with operant conditioning but I'm having trouble visualizing where to begin with this.

The following guidelines have proven to be useful when training macaques, including cynos, to cooperate during venipuncture for PK/TX [pharmacokinetic/toxicokinetic] studies.

A. General conditions
  › Avoid direct eye contact with the trainee; macaques interpret direct eye contact usually as a threat.
  › Always move in a slow manner, speak in a gentle tone of voice and use standard words/phrases.
  › Use small treats. If the animal persistently refuses to take a treat from your hand, don't embark on the training.
  › Avoid loud noises.
  › Sessions should be twice a day, after the morning feeding and before the afternoon feeding. Each training step is to be repeated until the animal is calm and cooperative BEFORE proceeding to the next step.
  › Remember, each animal is an individual and training should be tailored to the animals' responses.
  › Be patient; it takes as long as it takes.

B. Training steps
  1. Slowly restrict the animal to the front quarter of the cage using the squeeze-back. Do not immobilize the animal. Reassuringly talk to and praise the animal (e.g., "That's OK, Bob, good boy"). Scratch/stroke the animal through the bars and offer a treat. Release the squeeze-back and give a treat.
  2. Step #1 is repeated, but this time the animal should be positioned facing away from the trainer; the door is opened, just enough for the trainer to reach into the cage and scratch the animal on the back or thigh. The animal is rewarded with a treat and praised (e.g., "Good boy"). For security reasons, the animal always has to face away from the cage opening while being scratched. Release the squeeze-back and give a treat.
  3. Step #2 is repeated; this time the animal is only briefly scratched and one of his or

her legs is gently lifted and firmly pulled toward the cage opening. The animal is rewarded with a treat and praised. Release the squeeze-back and give a treat.

4. Step #3 is repeated; now the leg is gently pulled through the cage opening and stroked. The animal is rewarded with a treat and praised. Under no circumstances is the training session terminated before the animal's leg is successfully pulled through the cage opening for at least one minute. Release the squeeze-back, give a treat and praise the animal.

5. Step #4 is repeated and a blood sample taken by saphenous venipuncture. Release squeeze-back, give a treat and praise the animal.

6. Once the animal no longer shows any resistance, step #5 is repeated with the squeeze-back only pulled about 60%. The animal is now in control of the situation and has enough room to freely turn around, avoid being touched by the trainer or simply move away from the trainer. The animal is asked to present a leg (e.g., "Come on Bob, give me your leg"). If the animal refuses to cooperate, he or she is not punished in any manner but does not receive a treat. Release the squeeze-back but do not give a treat.

7. Step #6 is repeated until the animal cooperates and actively presents a leg behind or through the cage opening. The animal is rewarded with a treat and praised. Release the squeeze-back and reward the animal with praise and a treat.

Once the animal cooperates, step #7 should be repeated on a daily basis; during this routine exercise no blood samples are taken.

I followed these guidelines with minor modifications and successfully trained more than 50 rhesus macaques and eight stump-tailed macaques to actively cooperate during blood collection in their home cages. Once successfully trained, the macaques not only cooperated with me during blood collection but also with the attending animal care personnel.

It has been my experience that besides patience, mutual trust is the key to success when you want a macaque to work *with* you rather than against you. Therefore, before the start of the first training session I always spend an appropriate amount of time with the trainee to gain his or her trust and to feel safe in his or her presence. Once there is no trace of fear left in my relationship with the animal, the subsequent training becomes an easy-going interaction that we both enjoy. I don't hesitate to classify the training of macaques to cooperate during procedures as high quality environmental enrichment for the trainee and for the trainer.

I have not worked with any rhesus or stump-tailed macaque who stubbornly resisted during training sessions. They all reached the goal of the training program.

Does it take a long time to successfully train a macaque?

I took records while training 15 adult male rhesus and six adult female stump-tailed macaques. Cumulative times spent with an animal until active cooperation during blood collection was achieved ranged from 16 to 74 minutes, with a mean of 40 minutes for the rhesus males (Reinhardt, 1991); it ranged from 15 to 45 minutes, with a mean of 34 minutes for the stump-tailed females (Reinhardt & Cowley, 1992).

This insignificant time investment pays off in research data that are not influenced by stress reactions that macaques typically experience when they have not been trained to cooperate during blood collection.

Training macaques to cooperate during sedative injections

It is not uncommon that macaques are sedated via intramuscular drug injection on a regular basis for specific research-related procedures. Typically, the injection of the sedative triggers physiological stress reactions [Elvidge et al., 1976; Wickings & Nieschlag, 1980; Aidara et al., 1981; Streett & Jonas, 1982; Crockett et al., 2000; Bentson et al., 2003; Mori et al. 2006]. If the goal of sedation is the elimination/reduction of physiological stress responses to a particular procedure, the very act of sedative drug injection should not be stressful for the subject, otherwise subsequently collected data are under the influence of uncontrolled stress even before the procedure is being done with the subject.

Has anybody on the forum succeeded in training regularly sedated macaques to actively cooperate during drug injection in the subject's home cage—not restraint chair or restraint apparatus? It's not a big deal to train macaques to cooperate during non-consequential injections like daily insulin shots; injection of a sedative is probably a different story.

I have helped train many of our cynos to sit still for sedative injections at the front of their cages. For most animals we will pull the squeeze mechanism up halfway; applying clicker training, we first teach them to *sit* wherever they choose on the floor or on their perch. Some animals need the squeeze-back pulled up at about three-quarters, but they still have room to move away if they want. I have the impression that activating the squeeze mechanism acts as a signal for the animals that *it's time to work*. They respond promptly and come to the front section of the cage without actually being touched by the squeeze-back. Once they have chosen their location to sit, we say *arm* or *leg*, touch that body part, give the injection and reward the subject with a treat. Since they are going to be sedated, their treat consists of a piece of a Popsicle or cool pop. That way, even if the ice sits in their cheek pouch, it will be melted

PRIMATES 127

before they lose consciousness. We have a colony of only 52 cynos, but 45 sit calmly for us every time we have to inject them.

[I have trained quite a number of rhesus macaques to cooperate during saphenous venipuncture. Successfully trained animals required no additional training to also cooperate during intramuscular sedative injection; they simply presented their leg and kept equally still during injection as during venipuncture.]

Training macaques to cooperate during saliva collection

How do you teach/train macaques to RELIABLY allow you to obtain saliva samples for cortisol assessment?

The first step to training for consistent oral swabbing is to get a reliable *open* command. I work toward this goal using Prang in a curved-nozzle delivery bottle. Standing about a foot back from the cage front, I squeeze a light stream of fluid out (away from the animal's face, of course) and ask for *open*. I incorporate hand signals with ALL commands; for *open* I pinch the thumb to forefinger, and then open the circle while I vocalize the command *open*. At the first sign of opening, I click and reward. As the association of request to reward grows, I start raising my criteria, such as magnitude of open or duration of holding the mouth open.

Once you are getting a consistent open you will need to condition the introduction of an instrument into the mouth. I use a tongue

depressor. In the beginning I offer goodies like honey or yogurt on the stick to create a positive atmosphere towards the new item. It is important, however, to move away from that tactic as soon as possible in order not to cause confusion for the animal as to what should be done. If the animal continues to associate the tongue depressor with direct reward, he or she will continue to chew or eat the stick, which is obviously counterproductive.

Depending on the consistency with which you get your animals to approach and stay calm, you may need to condition for either a partial relaxed *squeeze* or a consistent *chest* to get close enough to swab. The best long-term scenario is to condition *chest* (calm presentation of chest at the front of the cage) and *steady* (holding this behavior/position until released by verbal praise).

When training for *chest* or *squeeze* I begin with light tactile conditioning. If an animal is a bit apprehensive, perhaps even fearful and potentially in a defensive mood, I use the tongue depressor for this, as it will keep my fingers safe and give the animal a positive sensation in association with the wooden object. I gently work my way up the trainee's body each day, from the more comfortable lower stomach region to the more guarded head, neck and finally, the cheek, chin and mouth areas.

Once I get a comfortable approach to the animal's mouth with the tongue depressor I begin reinforcing HEAVILY on even small successes. This is a potentially scary situation for the animal, so cooperation of any sort is to be commended!

If this step poses a hurdle, I resort to Prang in a delivery bottle; this will create a positive connection to the oral touching.

I find it helpful for the trainee and me to keep the duration of each session short. A session is discontinued immediately at the first sign of aggression, or at the first sign of discomfort, apprehension or stress. This gives the animal some control over the course of each session. Pushing the trainee beyond his or her limit can easily set you back weeks if not months; I am speaking from experience!

I usually do the actual procedure with a second person who reinforces the animal's cooperation while I carry out the swabbing. This keeps everyone in focus and safe, diverts the animal's attention and maintains a positive atmosphere.

My experience with training for saliva collection has been in the zoo environment.

Thanks for sharing this ethologically very well designed training protocol. In the profit-oriented research laboratory, time constraints may stand in the way of implementing it, which I personally feel is very unfortunate.

We did a study with two rhesus and seven cynos, assessing their cortisol responses to abrupt versus phased light changes. In order to avoid data-biasing stress reactions to blood collection via venipuncture, all nine macaques were trained to cooperate during saliva sample collection for the cortisol analysis.

We dipped the cotton ropes in mango juice, and practiced with each monkey, training them not to touch the rope with their hands, but to just chew/suck on the end of the ropes. Mango juice doesn't have the right pH for the actual test but the animals LOVE the mango flavor, so we used mango juice first to get them conditioned to readily chewing and sucking on the ropes. Once they learned that,

we stopped the mango juice and replaced it with coconut juice, which has the correct pH balance for the hormone test; we got that little adjustment down pat before we actually started the test.

The lab told me that a lot of times the saliva samples are compromised by insufficient volume OR traces of blood— which will ruin the test. If a macaque chews too hard or too long, he or she may experience a bit of gum bleeding. It's tricky to get them to chew/suck just long and gentle enough. The lab suggested rope chewing times of 2–3 minutes. I think that's why many colleagues end up with bloody samples. To get a better idea of that procedure, I chewed on one of the ropes to find out how long I had to chew to produce sufficient saliva; I noticed that the special ropes from the lab seem to suck the saliva right out of your mouth within just a few seconds. As a result of this little experiment, I had all our monkeys chew/suck the test rope for only about 10 seconds, and all samples were useable and free of microscopic traces of blood!

Sucking on flavored cotton ropes became a really fun enrichment activity around here; it certainly also helps with human-animal bonding. The monkeys love this! So do our interns working with them!

Are male macaques more difficult to train than females?

It seems to be commonly believed that adult male rhesus macaques are very aggressive animals and that chemical immobilization or mechanical or manual restraint is therefore necessary to protect the handling personnel.

If you have direct experience with the training of adult rhesus, cynomolgus or stump-tailed macaques of both sexes:

1. *Would you conclude that males are more difficult and dangerous to work with than females?*
2. *Would you conclude that it is more difficult to successfully train males than females?*

There is probably no black or white answer to your question because each animal has a unique personality. Let me demonstrate this with a few cases:

1. Holly, an adult female cyno is VERY naughty. Actually, Holly is more aggressive than any of the male cynos I have worked with. To train her would be a rather dangerous undertaking.
2. PearlySu is also an adult female cyno who is very sweet, shows absolutely no aggressive behaviors toward anyone or anything. PearlySu is extremely easy to train.
3. Justin, an adult male cyno is remarkably gentle, sweet and easy to work with, but not at all easy to train as he is not the brightest bulb on the tree.
4. Winslow, an adult male rhesus is aggressive but less dangerous than Holly; to train him is relatively easy.
5. Ivan, another adult rhesus is not dangerous; I call his reactions towards personnel *reasonable*. If he shows aggression, it is always under circumstances where I too would get angry. Ivan is a very sweet guy, easy to work with and not at all difficult to train.

I have always preferred to work with male rhesus, even though some of them were, indeed, potentially dangerous and would lash out and attempt to bite or scratch at the slightest disturbance. I felt a really good sense of accomplishment when they progressed during the training session; I had to be very careful, but they were intelligent and learned quickly.

I whole-heartedly agree with your thoughts about working with male rhesus. I also have a soft spot for the tough ones.

Having trained numerous adult rhesus macaques, I would definitely not say that males are more difficult to work with than females; but while training both sexes, I did get the impression that males learn the training steps more quickly than females. When I worked with males, I typically experienced that the trainee was really motivated to work with me, so the training progressed relatively smoothly and swiftly. Females had the tendency of being more hesitant, not so self-confident during the training; this often required numerous repetitions of training steps and hence more cumulative training time before the goal of the training was reached.

Treats as training tool for macaques

I would be very grateful for information that anyone could provide for me concerning the use of treats, such as sunflower seeds or small pieces of dried fruits, as an integral part of the training of macaques in preparation for experimental procedures.

Has anyone ever found the need to restrict the actual amount of treats used during training due to adverse or undesired side effects on the animal?

When training our cynos I have found that small treats like raisins, craisins, blueberries and peanuts (out of the shell) work really well as food rewards. Out of these four, the peanuts are my least favorite because peanuts tend to make the animals thirsty, a circumstance that can distract them. For this reason I tend to provide the peanuts as a finishing jackpot; they LOVE peanuts!

The raisins, craisins and blueberries are so small that the animals can receive quite a few of them as a reward (one at a time) during training without affecting their weight, appetite or health status in any way.

This is also my experience; when you use small treats one at a time as a reward during training sessions, you don't need to worry about any undesired effects with respect to body weight, appetite or health. I also use raisins, craisins, blueberries and peanuts in addition to PRIMA-Treats broken into quarters, Fruit Gems and popcorn.

I have used everything from certified monkey treats to Skittles and M&Ms. I try to stay away from sunflower seeds and peanuts because these treats require some processing, and that takes the animal's attention away from the training. Things that are small and can be eaten quickly are the best to use. Monkeys do have their preferences and some may want to work for gummy bears, others may prefer PRIMA-Treats. If you can find each monkey's favorite treat, it can influence the success of your training. I have not encountered any undesired effects using these treats as rewards during training sessions.

I agree; it is best to first find out what treats a particular monkey likes the most and then offer those as rewards during training sessions. I try to give the minimal amount of a treat that is sufficient to get the trainee to do what I request of him or her, and avoid treats that require so much processing/chewing that the trainee loses focus on the actual training.

The only words of caution I would offer is to reserve special treats for special behaviors; for example, a calm and collected allowing of an injection or other potentially noxious stimuli/event would be reinforced with a particularly favored treat. We use mostly banana chip pieces and raisins and reserve fresh fruit (grapes and apples) as special treats. Jicama and popcorn are the treats that we use for our overweight or diabetic animals. We do not use processed foods or candies as treats. For relatively long-duration behaviors, like blood draws or ultrasounds, juice is a good reinforcer.

We clear all of our reinforcement treats with the principal investigator. We have not had any problems with treats affecting scientific outcomes or health of the trainees.

I realized a long time ago that monkeys can have very strong food preferences. There was a rhesus male I worked with in a neurobiological lab who would only do his tasks for blue Skittles. As blue Skittles were hard to come by in large quantities, I tried to outsmart him one day and gave him blue M&Ms. He stubbornly refused to participate for the remainder of the session, and he almost looked betrayed. It became clear to me then and there that the least I could do was respect the animals' preferences and make the necessary extra efforts, if needed, to get the preferred treats for them. After all, these animals work for *our*, not for their benefit.

Training vervets to cooperate during procedures

Can anybody on the forum share experiences with the training of African Green monkeys (AGMs, vervets) during husbandry procedures (e.g., shifting, capture) and/or research-related procedures (e.g., injection)?

We clicker-trained some adult males and females but have not brought this to any productive conclusion yet in terms of blood sampling. However, we successfully made them enter various sections of their home and exercise cages, and enter into transport cages through target training. We found the greens to be quite receptive learners; it took about an hour per individual to obtain reliable cooperation. Some of them, including an entire family consisting of an adult female and male as well as their juvenile offspring, even learned tasks by merely observing what was happening in the neighbors' cage where the training was taking place.

Your very important observations suggest that vervets may, after all, be no less intelligent than macaques and, therefore, can also be trained to cooperate during various husbandry and research-related procedures. That so few training attempts have been documented may simply be related to the fact that the number of vervets in research is only a small fraction of the number of rhesus or cynos in research labs.

[I have observed vervets in their natural habitat in Africa and can confirm that these animals are pretty smart and learn not only how to steal your breakfast right in front of your eyes, but they can also figure out the usefulness of leaking water faucets.]

I found that greens were a bit of a challenge when we chair-trained them. They sat still and appeared calm in the chairs when I watched them. Once I turned my back, they struggled. I had a similar experience when introducing females to each other for pair-housing. Typically, partners began grooming each other immediately, but the moment I left the room, they started quarreling.

We successfully capture-trained and shift-trained over 250 socially housed, 1- to 25-year-old females and males for a cognitive task. Some of the females were also successfully trained to cooperate for periodic vaginal swabbing. Most of our caged vervets readily learned to jump directly into clean racks instead of being transferred via a transfer box.

Some people who work with rhesus say vervets are dumb. I strongly disagree but think that vervets are more anxious and/or more cautious, so they need a gentler *touch* in order to learn a task from a human. We've had stubborn animals, but as long as we remain calm they finally do come around.

I think if you have the right personality and calm demeanor and voice, you can train vervets to do almost anything. I always say *trust first*. Once they trust you, you're golden.

Training and behavioral pathologies in monkeys

Based on your own experience with monkeys, would you recommend formal positive reinforcement training [PRT] sessions as a therapy for animals who show behavioral pathologies such as hair-pulling or self-injurious biting?

While you are training an animal, the expression of behavioral pathologies are—most likely—suspended. My question refers to the expression of behavioral pathologies between training sessions when the animal does not get human attention. Do you find that monkeys on a training schedule show fewer behavioral pathologies even during periods of non-training?

During training sessions, behavioral pathologies are typically—but not necessarily—suspended; the sessions may, in fact, trigger self-injurious activities in some animals. I have worked with rhesus who exhibited self-directed biting while I pole-and-collar trained them. In those cases, I adjusted the length of each training session

PRIMATES 133

and stopped whenever the animal started to get tense; I gave lots of treats after each short session. With those individuals, I spread the daily training over several brief sessions. This greatly reduced the self-injurious biting reactions to the training, so it was worth being especially patient and spending quite a bit of extra time working with them. That is the least I can do for these animals!

I have noticed over and over again that our cynos express fewer behavioral pathologies when they are on a training schedule. We have one male named Rock who would constantly scratch and bite at his testicles, fingers, arms and legs. He was doing a lot of damage to himself; we found him a compatible cage companion, and this greatly diminished his pathological behavior. However, when his partner was on study and the two needed to be separated for the day, Rock would go back to hurting himself. We tried providing foraging devices and puzzles to keep him busy, but none of it worked for very long.

Once we began training Rock for pole-collar-and-chair whenever his partner was on study, he stopped hurting himself. We now no longer train him every time his partner is on study but instead have varied the training schedule to every other time or every third time his partner is on study. He is always ready to work with the trainer and really seems to enjoy sitting in the chair and getting attention.

To our great relief, Rock no longer hurts himself when he is periodically separated from his partner during the day, but a full day plus night separation from his companion turned out to be too distressing for him and he would invariably start biting himself

during the night. So we do make sure that he and his partner are always paired overnight. Everyone who works with our cynos knows that Rock and Ray are the exception to the rule; they are always pair-housed during the night, no matter what. This extra attention to Rock's special need for social companionship is worth all the effort: Rock has become a behaviorally healthy animal and no longer engages in self-injurious biting.

Back when I was doing neuro research, I had a rhesus who saluted on a regular basis—even when he was on the pole and being walked to the scale; he would actually sit on the scale in order to salute! But, when he was working and directly interacting with a person, the saluting all but disappeared. After about six months of a regular testing schedule, the animal hardly ever displayed the bizarre saluting gesture on days when he was *not* working, but it still occurred occasionally. His cage mate however, displayed aggressive mouthing whenever he felt threatened or stressed. I attempted to train him as well but, sadly, his mouthing would exacerbate into SIB [self-injurious biting] during the first training sessions. We opted not to use him as a research subject and stopped the training, but kept him as a social partner.

In that same lab, I had another animal who did back flips for what seemed like the entire day until I started to pole-collar-and-chair train him. After about a week, I noticed that the back flipping started to decrease; and then in the seventh week—right at the time when formal testing started—the behavioral stereotypy stopped altogether.

Making use of quarantine time

When you are in charge of new monkeys while they are in quarantine, do you take some time to familiarize the animals with you, perhaps even try to develop a mutual trust relationship with them and/or make them familiar with the clicker before they go on a study?

I currently work in a biomedical facility that doesn't have many long-term monkeys, but we do take full advantage of the quarantine time to desensitize the animals to the presence of humans and to habituate them to routine procedures. This helps them to adjust to their new environment and it allows our staff to get familiar with the individual monkeys well before they go on study. The development of such relationships helps to foster mutual trust, which can then buffer possible fear responses during procedures.

Our SOP [standard operating procedure] is geared toward building a positive relationship with new macaques in quarantine in order to help the animals feel relaxed in the presence of personnel and get acclimated to basic handling procedures. Food treat rewards play an important role in this process.

We have seen a big difference in the animals when the extra couple of minutes per day are invested; at the end of the quarantine time, the animals are much calmer and come forward to the cage front to get the person's attention and then be rewarded with a favored treat.

We work mostly with cynomolgus monkeys. The rhesus tend to be more nervous and apprehensive, so they need more attention and patience before they settle in and get more comfortable when a person is present. But once we make friends with them, the extra effort pays off in friendly animals who willingly work with us.

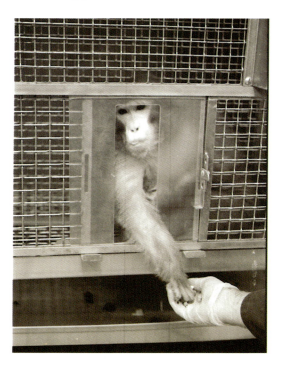

It has also been my experience with macaques that this extra time spent with new animals has quite a number of important benefits: (1) high quality mental and social enrichment for the new, isolated monkey, (2) high-quality enrichment for me, (3) safer and easier handling of the animal during the study, and (4) valuable foundation for any formal PRT project with that animal.

One of the most rewarding things we have done is housing new arrivals in compatible pairs as soon as possible; we do research with rhesus and cynomolgus macaques. Our facility is small enough so that we can quarantine monkeys in all-in/all-out fashion. It didn't seem rational from a disease exclusion standpoint to individually house the animals when we could house them in pairs or groups. Because New York State has strict quarantine requirements even for macaques who come from within the U.S., we obtained special permission for this practice from the Department of Health, recognizing how diseases are transmitted and the important animal welfare implications.

Although we haven't scored it objectively, attending staff report that monkeys in quarantine are now doing much better, especially when subadults are brought into our facility where they are housed in pairs rather than alone. They almost immediately change their response from being terrified at the far corner of the cage to co-threatening with their cage mate against personnel looking at them. In general, they seem so much less apprehensive, even during TB testing. We have only recently begun this practice, with dedicated staff who also condition the animals to remain calm during husbandry procedures. For sure, research techs, vet techs, and animal care staff have noticed a positive difference.

If your facility can get past disease exclusion issues, pair- or group-housing monkeys in quarantine is definitely the way to go.

Many years ago I started pairing and habituating all our new rhesus arrivals while they were in quarantine. It didn't take long for people to realize that the animals who had positive experiences through quarantine were easier to handle once they were on study. Pair-housing macaques soon after arrival is rewarding, and I believe, much better for the monkeys and for the staff; it is great!

I work in a pretty fast-paced CRO [contract research organization] where we have quite a few quarantine rooms open concurrently. Our facility primarily houses cynos.

With new animals in quarantine I start with a simple exercise. The animals are pair-housed, and I target them to their feeder for cooperative eating. As trust is established, I work up to the eventual goal of restraining arms through the doors. The majority of our procedures are based on the animals presenting arms for restraint. The ultimate goal is to provide a completely trained monkey by the end of the quarantine period. I have yet to reach that goal in the five weeks of quarantine, but I think it's feasible.

When we get in a new room of 16 monks [rhesus or cynomolgus macaques] in quarantine, we start working with the animals to build trust, having them come to the front of the cage and accept a treat. This can turn into a vet check at cage side, especially if you can prompt them to present various body parts. Then, when they come out of quarantine, they have some trust with us already built, and this helps when we train them for pole-and-collar and then for chair restraint.

The attending care staff plays a very important role when you want to have new arrivals prepared for research while they are in quarantine. Right now I have fantastic care techs who really enjoy being with these animals, so they spend extra time with them. This helps tremendously! The monks seem happier as well. They quickly learn to trust humans, which in turn is the basis of training them to work *with* humans. When I go in the room, I can see the difference it makes for those monkeys in quarantine, and how fast they learn to cooperate with the vet staff. I think it's really important to have care staff who truly love monkeys and enjoy working for and with them.

Behavioral pathologies in macaques

Hair pulling [overgrooming, hair plucking, trichotillomania] resulting in alopecia in monkeys and apes is traditionally treated rather than prevented.

Have any of you successfully treated behavior-induced alopecia in non-human primates to the point that the subjects had their hair regrow to normal coats? What kind of treatment did you apply?

We had an adult male cyno with severe hair plucking of the face, head and arms. You could observe him performing this behavior, and his wincing proved how uncomfortable it was, but no amount of enrichment or training seemed to help this male to stop pulling his hair.

We recently moved into a new building where we had to combine some rooms and separate others; this resulted in one female cyno being moved into this male's room. The male's behavior changed almost immediately with the introduction of the female; the hair plucking stopped completely. We were worried the hair wouldn't grow back, but after almost a year, you can barely distinguish him from the other males. It's amazing what boys will do to impress a girl!

I have worked with a rhesus male who was almost bald as a result of hair pulling. He was on a long-term study requiring that he be sedated every two weeks for VAP [venous access port] maintenance. At one point the VAP failed and the male was released from the study. Once we were no longer sedating him every two weeks, his hair started to grow back, and after a few weeks you couldn't guess he was the same monk who had plucked almost all his hair; he was a very sensitive guy. I think the repeated stress associated with sedation was the cause his hair pulling.

From the limited exposure I've had with our pig-tailed macaques, I did notice that alopecia (mainly arms, lower flanks) did become much more noticeable when we had to separate long-term-paired boys.

These observations suggest that compulsive hair pulling—like self-injurious biting—can develop not only as a result of species-inappropriate raising and housing conditions but can also be triggered by stressful situations, such as repeated sedation or separation of cage mates.

I run a small breeding colony of cynos for reproductive toxicology studies and just recently have had three of my male breeders start to self-bite. I've noticed that it happens when these males haven't been used for a while. They share the room with other males who are breeding females in the same room.

Does anyone have any suggestions regarding the reason for the self-biting behavior of our males and how we could prevent it?

For sure, the males watching other males with a female could start problems—sexual frustration among the males without girls could be the cause. Separate breeding areas would prevent this and probably eliminate the cause that triggers self-biting in some of your males.

My best guess is that a male who is singly housed, but can see another male courting and copulating with a female, can get extremely frustrated. Not being able to release either his prompted sexual drive or his prompted aggression drive, he resorts to an aggressive substitute behavior, thereby getting at least some of his tension released. If you must keep the breeding paradigm this way— and I'm willing to bet you need to, due to the toxicology aspect—I have a suggestion: one of the best things I have found is to line up the cages of males on opposite sides of the room

and place an opaque curtain—shower curtains are good because they can be easily hosed down—in the center of the room that can be closed when breeding takes place. That way, those males who are not actively breeding won't have direct eye contact with those who are. They'll still be able to smell and hear one another, but the lack of full eye contact with a breeding pair prevents excessive aggressive and sexual arousal. Also, try to ensure that the guys that aren't breeding are *over*-enriched in some way, and give them something, like a large Kong toy or a big branch segment, that they can bite on. These small modifications could Band-Aid your situation.

We used this approach in a room filled with 24 male rhesus breeders who lived in 12 compatible pairs: two rows of six pairs on one side of the room, separated by an opaque shower curtain from two rows of six pairs on the other side of the room. Needless to say, the partners of each male pair were temporarily separated by a blind double-cage divider when a female was introduced to one of them during an approximately 24-hour breeding period.

With this housing arrangement, self-biting was never witnessed or retrospectively reported in any of these breeders during the 10 years I worked at that facility.

I did see self-injurious biting in single-caged males, and fighting between previously compatible males who were accidentally exposed to the sight of male-female breeding pairs; I think this is not really surprising.

We have discussed on several occasions how transfer to compatible social housing arrangements can bring injurious self-biting to an end in macaques (Baumans et al., 2007). Self-biting is not uncommon in single-caged baboons. How do you treat this behavioral pathology in baboons? Does foraging enrichment or inanimate enrichment make any difference, or do you also have to find a way to provide compatible social enrichment?

We found that while access to a foraging log had no curative effect on serious self-injurious biting of a single-caged subadult male baboon, transferring the animal to an outdoor section with compatible females in adjacent cages (allowing grooming interactions) resulted in a healing of the self-inflicted laceration within four months. After 18 months neither the self-injurious biting nor the wounds re-occurred (De Villiers & Seier, 2010).

It's not really clear which of the two variables—living with females and/or living outdoors—healed the baboon of SIB, but I would assume that physical contact/interaction with the females was the key factor.

[Crockett & Gough (2002) noticed that an adult self-biting female baboon stopped biting her knee once she had started to make a Kong toy, rather than her knee, the target of tension-related aggression.]

Has anyone had issues with monkeys licking or eating their feces? I've one rhesus who predictably shoves feces into his mouth before being taken out of his cage. Now I believe I have a cyno who is smearing feces on the bars of his cage and then licking it off. It doesn't appear to be biscuit mush, but I haven't actually seen him pick up the feces and rub it on the bars either.

Some macaques will indeed play Picasso with their feces—something on which to expend their creative energy, so to speak. However, if he's actually ingesting the fecal matter, I would check if he's suffering from a dietary problem such as not enough biscuits or a vitamin deficiency of some sort. I've cared for a couple of both rhesus and cynos who have been fecal eaters; once we either increased the biscuit ration or started them on vitamins, the abnormal behavior stopped for good.

I have one rhesus who shows this weird behavior with consistency the day after he has received an orange. He kind of tries to re-process the orange the following day, smearing his orange-colored poop all over his face and picking out pieces of orange pulp and eating them.

I saw feces eating regularly in chimps, many years ago. These animals were kept in conditions that I hope no one would permit nowadays. They were also very efficient at hurling feces at researchers—never at the techs—with deadly accuracy, so I am convinced that the feces-oriented behavior was stress related, probably triggered by fear-inducing circumstances.

We have a rhesus who not only was a voracious eater but he also ingested his feces. He was diagnosed with diabetes; insulin therapy brought the eating problem under control and stopped the consumption of feces. You might rule out diabetes before deciding it's a behavioral problem; just in case.

Touching non-human primates

At your facility, do you have a no-touch policy with non-human primates?

If you are allowed to touch the animals, are there circumstances where you feel that the animal truly likes it when you touch or groom him or her? Do you find that one gender is more receptive to human touch than the other gender?

We never had a strict policy on touch, but handling animals who are not anesthetized is always a concern. Non-human primates are not tame/domesticated by any means. They may be small in stature but they sure could do a lot of damage to a person [who is putting them into a situation that calls for self-defense]. Despite not being overly aggressive animals, rhesus monkeys do have sharp teeth that can transmit viruses.

Being touched by a human is something each animal would need to adjust to. Depending on the touching person and depending on the monkey's experience with that person and with humans in general, some animals may like to be touched and others may try to avoid it. We had one rhesus who loved to be held and picked up—through the cage bars—by one particular technician, but she would never allow me to do that, nor would I have felt safe doing it. There was a trust-bond established between this particular animal and the technician.

In my opinion, safety ought to be the primary concern—mostly for the person who interacts with the monkey. But I do think that non-human primates would benefit from human touch, because being social creatures they are likely to enjoy being touched by another social creature whom they can trust. I would think that being touched by a well-meaning human can make a monkey feel less fearful and perhaps more comfortable during procedures that require direct contact with a person.

My facility also does not have a no-touch policy. Since safety is our primary concern we do discourage our staff from just going around, trying to groom monkeys. All our animals are male cynos. They let you know quite clearly whether or not they prefer to be left alone or to be groomed by you.

    We have one male who responds to women very differently than to men. He (Junior) allows only women—me, an investigator and two of our vet techs—to groom him through the bars of his cage or while he is in his chair during training. Junior is biased against men and does not want to be groomed by any of our male animal care employees. He invites us women by placing himself up against the bars of his cage and closing his eyes and then rotates himself around to make sure we get all of his favorite spots. He usually falls asleep when one of us is grooming him gently, but he does not like it when any of the men try to do the same.

    I believe that being touched and groomed by accepted and trusted humans is VERY enriching for Junior and several other monkeys of our facility, who behave in a similar manner as Junior does. I am SURE these particular monkeys like it when they are touched by certain—not all!—people.

Your story reminds me of Tadatoshi's male Japanese macaque who had a clear preference for movies with women (Baumans et al., 2007; Ogura & Tanaka, 2008).

    These observations suggest that gender does play a role in the calming/distressing effect human caregivers have on the animals in their charge; this topic has yet to receive attention in the published literature. It may be an important, yet overlooked extraneous variable in certain cases.

We do have a no-contact policy at our facility, as safety must always come first. Any unauthorized contact would result in immediate dismissal. However, there are staff members who have extensive experience and have been working with our monkeys for 5–13 years; they have developed affectionate bonds with certain monkeys and understand their unique personalities and can read them well. We do make exceptions to the rule for these staff members, but we also have a few monkeys that nobody is allowed to touch for strict safety reasons. One monkey stands out in my mind: Holly; she is our brat. Holly is a pincher and has developed her trick to perfection. She causes more damage pinching than others can cause with a bite. She enjoys it! She's clever and will lure her victims by deception to get them close enough to accomplish her mission. NO ONE is allowed within arms' reach of this macaque brat.

    Our pair-housed monkeys don't really need and they would probably not benefit that much from human contact, as they have a compatible companion to provide the necessary elements of touch.

    We also have a few monkeys who have to be singly housed; these animals need direct human contact, they beg for and they benefit from human touch. Some of these individuals arrived with serious SIB and other neurotic behaviors. They had been singly housed for years without ever being touched by a monkey or by a person. In a short time, the touch they routinely receive from our seasoned staff is deeply affecting their emotional well-being, and many of their neurotic behaviors all but disappear. These rescues pose in different positions to receive their *therapy* in just the right spot, and they will fall asleep while being groomed. For them, human touch has a truly therapeutic, healing effect.

Our facility does not have a no-touch policy specifically, but we advise against touching the monks for safety reasons. Who wants to have to deal with avoidable occupational health issues!?

Some of our rhesus macaques—both females and males—seem to really enjoy being groomed by humans. They will approach the front of the cage and present different body parts for you to groom and then rotate so that you move on grooming another part of their body. They get that very typical glazed, blissed-out look—eyes at half-mast and completely relaxed body. This kind of human-animal interaction is enriching for both parties involved, the animal and the tech who does the grooming. Some of the rhesus males like to be scratched on their heads and necks while sitting in their restraint chairs with blissed-out faces. I worked with a large male who, while sitting in the restraint chair, would rather have me scratch his rear and hips as a reward than getting juice or treats. Similarly, a younger cyno female had me regularly hold her hand while chaired for blood draws. But

I have also had some real *brats*, as Polly so aptly puts it. These little devils are so great at luring you in and then grabbing and pinching whatever they can get hold of you. Touching them, let alone attempting to groom them, wouldn't be a good idea!

From what I have experienced with monkeys, I don't think there is a sex difference in the monkey's bias for female versus male humans or in the monkeys' receptivity to human touch. The one caveat to that is: females—mainly rhesus—while they are cycling can be a bit more cranky, so they usually tend to be less receptive to human touch.

At our facility there is no rule that you can't touch a monkey [rhesus]—in fact we train them to present their legs to us while they are sitting in the restraint chair, so that we can touch and hold their legs as we place IV catheters; for the grabby guys we do place shields up, but only if they make an attempt to reach for arms, hands and face with the intention of scratching or grabbing.

There are several females and males of various ages at my facility who will seek out our attention from within the cage and solicit grooming. Some will place their entire arm out of the cage and let you groom them for several minutes, while others press their chest or hind legs against the mesh of the cage front hoping that you touch or, even better, groom them. I believe these particular animals enjoy being groomed by humans as they (a) seek it out, (b) relax as the grooming goes on, and (c) will start grooming my jacket or latex gloves after I've groomed them for a while.

We do not have a written *do not touch* policy. Much like everyone else, we do not advocate touching, for safety's sake.

At all the places where I have worked—my current included—there were moments when I felt it is just fine to reach out and touch. I've had cynos, rhesus, pigtails and boons of all ages and both genders present in an unmistakable invitation gesture for a good scratch. Currently I have two aged rhesus males who tend to fight for my attention during rounds. The dominant one will put his rear end up against the cage bars, and when I approach will settle down so I can give his entire back a good scratch/groom. The subordinate one will see me in this act and will then saddle up to the front of the cage and present his body so that I can scratch his chest/belly/face, whatever—he's even presented his tongue to me. When either of them is really happy, they'll go into a kind of trance state; the subordinate guy has even sighed a couple of times. Then there are the days that, once I get done with these two, the rest of the room—all aged rhesus males—begin to present body areas for affection. Granted, we also have our brats as the others have described. I refer to them as teases: butt is fully in the air, but as soon as you approach and reach out, they perform that lightning fast turn to give you some attitude.

Our cynos can be just as affectionate. We have one male who will demonstratively lie down in his cage, fixing you with his eyes to see if you're willing to get close enough to groom his belly, at which point he will roll over for more exposure—he's quite the character! But, he also has his cranky days as well. So, when staff members see me grooming any of our critters and comment on how calm they are, I explain that not all monkeys are this way, and that even these calm animals aren't this friendly all the time. Like people, they have their bad days as well, and you better respect this and leave them alone!

I agree there isn't a difference between males and females in regard to affection toward humans. I also agree that cycling females can be somewhat picky as to whom they like or dislike. One female pigtail stands out in my memory. She was on a viral tox study, and didn't get along with almost anybody on the staff. She wouldn't show any type of affiliate behavior or gestures towards people in general and would even lash out from time to time; but for some reason she had taken a shine to me. I could never figure out why, but whenever I entered the room, she would stand, hoot, and duckbill in my direction until she got my attention. Then, when I would walk to her cage, she would calmly settle and present her hips to me for grooming. We got to a point where she would reach out of her cage and attempt to groom me. At this point, I started to don extra PPE [personal protective equipment] so she would have something to groom. After a couple of months, she started to calm down a great deal and wasn't as cranky around other people, which we all felt was in everyone's best interest. She would always be quite standoffish right before her cycle, but as soon as her *beanbag chair* started to deflate, she would return to her old self.

Experiencing these affectionate relationships with animals, who have ample reasons to be suspicious of people, are highlights in our daily work. It is really amazing how sensitive these critters are and spontaneously distinguish a person who genuinely cares for their well-being from another person who does not feel emotionally connected with them. Attending personnel and investigators are usually not considered as extraneous variables that can affect the research subject's physiology; I think this variable can have a significant impact on research data and, therefore, deserves more attention.

At our facility no unprotected hand may be placed into a cage in which a non-human primate (NHP) is kept. Appropriate PPE must be worn when handling/touching an animal, and when in doubt, keep away.

New staff members are mentored when working with NHPs and they quickly learn from the animals' behavioral responses which individuals are okay to be cage-side groomed and which are not. Those who eagerly accept a treat are generally also receptive to being groomed through the mesh of the cage front, while those who threaten when being approached are better left alone.

With some of our animal training protocols, visible acceptance of human touch is then positively reinforced to finally obtain the monkey's cooperation during procedures such as cage-side vaginal swabbing.

I do find our cyno females much more receptive to touch than the males. My view may be a little biased as we have many more opportunities to actually touch females than males during procedures.

When I worked with single-caged baboons it was obvious that these animals did not really like to be touched, let alone groomed. They accepted food from you but showed no further interest in any other interactions. There was one exception to this rule: the female baboons became more receptive when they were in estrus and some would actually present their bums as an invitation to be firmly scratched; once the estrous phase was over, they no longer tried to get human attention.

MISCELLANEOUS

Dealing with repetitive locomotion/movement patterns

The implementation of environmental enrichment is often prompted by the fact that confined animals engage in repetitive locomotion patterns; the enrichment is then supposed to reduce or eliminate these so-called undesirable abnormal activities. As long as stereotypical movements (e.g., running in circles, back-flipping, walking back and forth) are not harmful to the individual animal, do we really need to be concerned about them?

The stereotypical behaviors of pacing, rocking, flipping, etc., while possibly not harmful in and of themselves, are an indication of a deeper problem such as boredom, frustration or distress. It is beneficial for the animal's well-being if we can find ways to alleviate the root cause, such as with enrichment, appropriate rearing conditions, and social interaction—thereby removing the environmental triggers of these stereotypies in the first place.

I worked with 5,000 rhesus macaques in a lab setting. The stereotypic, repetitive patterns were almost always shown by singly housed animals, not by group-housed animals who had access to well-structured outdoor quarters.

MISCELLANEOUS 145

For me, it's less a question of what confined animals are doing than why they are doing it. While repetitive movements may not be physically harmful, is there an environmental stressor that prompts the behavior?

We have some strains of mice at our facility who run in a vertical circle in their cages when there is disruption in the room. The disruption can be something as common as a room health check where someone has to look into the cages, or jarring, e.g., when somebody is dropping a stack of cages. While the acrobatic maneuver is clearly not something that's going to injure the mice, I'm always concerned about overlooked ramifications of stress triggers.

When stressed, I personally have a habit of clicking my writing pen on and off. I'm not aware that I'm doing it and it's certainly not hurting me to do that motion. But is the stress I'm experiencing harming me in other ways? What about people near me who are bothered by the constant clicking sound? What impact is that having on them? Those are the same questions I have for the animals in my care. If the stressor that triggers stereotypies is something that we can avoid, eliminate or minimize, then I think it's worth pursuing.

We do know the root cause of common stereotypic locomotion patterns, such as running in circles or pacing: insufficient space. Unfortunately, knowing the root cause does not necessarily help us deal with some of the basic husbandry-inherent problems—such as too small cages—because of the financial burden that would accrue from the correction of the problem.

As a veterinarian I am not really concerned about these locomotion stereotypies, but as an ethologist I am very concerned about the inadequacy of the animal's living quarters and try to adapt them to the animal's spatial needs for species-typical locomotion. When this is possible, other animals who are raised and kept in such refined/improved living quarters will no longer have any reason to develop stereotypical locomotion and movement patterns. When we prevent stereotypies from developing, we no longer need to treat them—which, as we all know is a lost battle in most cases.

I agree, prevention of stereotypic behaviors—rather than waiting until they manifest and then try to treat them—is the best way of dealing with them.

People looking at these repetitive behavior patterns typically know at a visceral level that all is not well with the individual animal who is engrossed in them. Excusing these activities as not harmful or as a normal coping strategy distracts from the fact that they do indicate that the animal is in distress.

Personally, I think we are wasting our time and resources when we try to treat only the symptoms—repetitive locomotion and movement patterns—of this confinement-created problem. Temporarily distracting a rat, a dog, a monkey or any other animal from engaging in stereotypical locomotions, or even making it impossible for him/her to engage in them can be easily accomplished, but does it really help the spatially frustrated animal, and does it actually address the cause of the problem? After all, the stereotypical locomotion only shows that the animal is desperately trying to somehow cope with enforced confinement in poor living quarters. If we listen to what the caged animal tries to communicate, we can get to the root of the dilemma and design better, more species-adequate living quarters in which animals no longer have any reason to engage in activities that we humans label as undesirable or abnormal. What is undesirable or abnormal are not the animal's behavioral adjustment attempts, but the human-enforced, badly designed living quarters.

When I observed single-caged rhesus macaques stereotypically running in neat circles and contorted figures of 8, somersaulting, bouncing and/or moving the body in a swinging motion from one wall to the other [often accompanied by flapping a hand against the wall], I got the impression in many cases that the subjects got relief from unspecified tension, perhaps even from distress resulting from frustration and chronic boredom. It was obvious to me what was wrong—not with the macaques, but with the housing arrangement:

1. insufficient space,
2. no spatial structures,
3. no social companion.

The solution for these behavioral, human-created problems was easy: I paired all these animals with compatible conspecifics and moved the pairs into double-sized cages, each section furnished with a high perch.

The stereotypies came to an end within a very short time. I am sure they would have reappeared if the animals had been transferred back into unstructured single-cages. I believe that stereotypical locomotions [and probably all other stereotypies, including trichotillomania and self-biting] could be eliminated in macaques if all animals were raised and subsequently kept in:

1. spacious enclosures,
2. properly structured enclosures, and
3. compatible social settings.

I would not classify these three elements of the animals' living quarters as generous enrichments but as basic necessities.

The individual variation in the occurrence of stereotypical movement patterns supports the idea that there are monkeys suited for laboratory environments versus monkeys who should not be housed in a cage. Is selective breeding of hardier (in the sense of being more cage-tolerant) individuals a tool that can be used to combat the occurrence of unwanted behaviors?

This way of thinking reminds me of the company that bred hens to be blind so they would not feather peck: the undesired behavior disappeared so the behavioral problem was solved, but at what moral and ethical cost?

Addressing social needs of animals who have a bandage

When a socially housed animal has a bandage or a cast as a result of an experimental or medical procedure, how do you address the individual animal's social needs? Do you take the risk and let the animal share a cage/pen with her or his compatible companion(s), or do you or other staff members substitute for the conspecific companion(s) until the bandage or cast has been removed?

MISCELLANEOUS 147

We have tried to house pigs together after surgery but realized that they can't leave another fellow's bandage alone. They like to chew and pick at anything, especially something unusual that needs to be explored.

My solution is to single-house the pigs post-surgery, but I spend a half hour or so each day with them (depending on the number of pigs). They seem to enjoy my company very much, especially when I come along with a brush and give them a good massage.

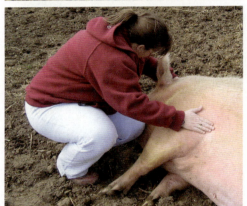

I have housed pigs together with a companion post-procedurally with a stocking net over the actual bandages. The net may get nibbled, but the animals mostly leave it alone, as they mainly want the contact with the other pig.

The trick with tight netting material is a great idea! I would assume that it works with pretty much all warm-blooded species commonly found in research labs.

I can say it works with cats in the home environment as well. For a time I had a sizeable number of cats all under one roof. One of them managed to slice his foot open and five of the other cats kept trying to pull the vet wrap off his foot. I covered the wrap with stocking net and made a few small balls of vet wrap that I tossed around the house. They'd all play with those balls for hours but left his stocking net alone for the 12 days it had to be worn.

In the research setting, I'm definitely a fan of keeping animals together unless the protocol really does require that they need to be isolated. Obviously, it's easier to separate companions after procedures, so that's what happens all too often.

Squirrel monkeys are experts in picking at bandages on themselves and other group members. I remember reading somewhere that they don't like it when others look different than usual, and will show aggression against such individuals. We had such a situation and had to separate a previously compatible pair.

We re-socialize our mongrel dogs usually the day after surgery; so far we haven't had any major issues with bandages being chewed off. My personal observations are that the dogs perk-up quicker in the company of their buddy than being left alone to recover, but there can be exceptions, especially if your buddy is an excessive mounter.

Radio sounds in animal rooms

Being exposed to CD or radio sounds may be a nuisance (stressor) for some human primates, while others get entertained by it. Animals are often exposed to this kind of noise, which serves as environmental enrichment for attending care staff, so I am wondering what effect does it have on the animals? Do they also enjoy the music and human talk, or would they prefer to have silence?

We play radio for our rhesus macaques three times a week. I honestly have never noticed a difference in the animals' behavior when the radio is on or when it is switched off. We play the station that my co-worker prefers—70s and 80s rock; we are in the basement, and this is virtually all we can pick up. I am in the process of getting new radios that can

play audiotapes/CDs, so I can offer classical music and nature sounds—with the monkeys' interest in mind.

As far as the other critters are concerned, I can share thoughts on rabbits. I think they would prefer silence to human-created sounds. However, the sound of music may diffuse all the sudden noises that humans make when they are in the room; these noises are probably more disturbing for the animals than radio music. But I believe that the radio should always be turned off when humans are not present so that the rabbits can peacefully relax.

Music affects the mood of our rhesus and cynomolgus macaques quite a bit. Gregorian chant music, Indian flute music, and soft drumming have a noticeably relaxing effect on all our monkeys. When we turn the TV to the music channels on occasion, we always put it on a channel called Beautiful Instrumentals; that also has a calming effect.

Our macaques do not like loud, continued music at all, even music that normally relaxes them. We always keep it at a volume that would allow us to fall asleep.

Hard rock and heavy metal music is extremely irritating to our animals, so we never play that type of noise at any volume.

I would love to develop a device that allows our macaques to turn the radio on and off at their own will.

Line et al. (1991) exposed ten adult rhesus macaques to such a radio—preset to a soft rock format station at a low volume—for a 6-week test period. The animals could turn the radio on and off by touching contact detectors extending from the apparatus into their cages. Some monkeys never turned the radio off while others never turned it on. The radio was used with an average playing time of 1.3 hours per 24-hour day (Line et al., 1990).

Every day, an audio system plays two hours of music and nature sounds in each room of our rhesus and cynos. It is remarkable how relatively quiet the monkey rooms are when the audio system is turned on.

We recently started playing music for our pigs. Already I have noticed a big difference. The rooms that have the music playing seem so calm, all pigs are quietly lying in body contact with each other. They barely move when I come in the room. They all appear so relaxed; it is great!

Kilcullen-Steiner & Mitchell (2001) "effectively decreased the amount and intensity of barking" by exposing dogs to white noise along with new age music. Wells et al. (2002) observed dogs housed in a rescue shelter and noticed that "classical music resulted in dogs spending significantly more of the time quiet than did other types of auditory stimulation" such as human conversation, heavy metal music and pop music.

[It has been shown not only in human (Tse et al., 2005; Zhou et al., 2011) and non-human primates (Brent & Weaver, 1996) but also in rats (Sutoo & Akiyama, 2004; Akiyama & Sutoo, 2011) and mice (Hu et al., 2007; Núñez et al., 2007) that exposure to gentle music— not noise—can have a significant stress-buffering, health-enhancing effect.]

We have radios on in all rodent and rabbit rooms, and we certainly find that it buffers the startle factor.

It is a bit surprising that so little research has been done to examine if the sounds/ noises that are produced by the radio are actually preferred over silence by the animals who, after all, have little choice in the matter, as they are confined within the room. Some animals, especially nocturnal animals such as rats and mice, may feel disturbed, perhaps even distressed when being exposed to these sounds/noises. This could be a concern for investigators who want to make sure that their animals are not exposed to extraneous variables that might affect research data.

Checking the literature, I find only two preference studies, one conducted with rats, the other with marmosets and tamarins. Both studies found that the animals prefer silence over music/radio:

1. Krohn et al. (2011) exposed rats to different kinds of sound patterns, including radio, and found that the animals "showed a clear preference for silence to anything else, which may be taken as an indication that they feel disturbed by the sound from the speaker."
2. McDermott & Hauser (2007) showed that "both tamarins and marmosets preferred slow tempo to fast tempo music, and when allowed to choose between slow tempo musical stimuli and silence they preferred silence."

More such preference studies are warranted to check objectively if the animals in our charge prefer the radio music/talk that makes us feel good, over silence. If they prefer silence, we should respect this as long as the animals cannot turn the radio off themselves. We all try to give our animals some control over their environment as part of enrichment programs; if we use the radio in the animal quarters it should, in my personal opinion, be such a controllable element of their environment.

Phased lighting in animal rooms

Does anyone have references supporting the use of phased lighting, simulating sunrise and sunset in monkey rooms? It seems to me the abrupt change of the lights going on and off might be stressful for the monkeys.

I can answer this from the perspective of common marmosets.

We use a 12-hour light-dark cycle, with the animal room lights gradually coming on and then going back off in a series of steps, to simulate dusk and dawn periods.

When there has been a need for me to stay in the animal rooms until evening, I have noticed that when the first room lights switched off, the animals became subdued and quiet; by the time the second set of lights switched off, the animals had begun to retreat to their nest-boxes, so that by the time total darkness had descended, the animals had sufficient time to close up shop and be settled in their nest-boxes for the night. If the lights would go off all at the same time, many of these animals would probably spend the entire twelve hours of darkness on the floor or on the perches of their enclosures, not in their nest-box.

I would strongly recommend simulated dusk and dawn periods as an integral part of the light cycle for marmosets to foster their welfare in the laboratory setting.

In commercial free-range hen flocks, it is common to give the hens an artificial dusk and dawn in the house. The dusk is particularly important as it allows the birds to fly up to roosts while there is light available—thereby avoiding injury—rather than in total darkness after all the lights have gone off. The dawn and dusk are usually achieved by turning one third of the lights on/off, followed by the second third approximately 15 minutes later, and finally all the lights on/off another 15 minutes later.

We have a staggered light system for our rhesus macaques. At 6 a.m. only half of the lights are turned on. Shortly before our technicians enter the room to do their morning checks, they turn on the remaining lights. In the late afternoon the techs turn off half of the lights after they have checked the animals, and they turn off the remaining half of the lights at 6 p.m. We also have large windows on our doors, so there is some light filtering in after lights out. I don't have any info on the impact on the animals, but it is a relatively simple procedure that can avoid the drastic light change.

We recently conducted a study that compared dawn and dusk phases; the research subjects were two rhesus and seven cynomolgus macaques.

We observed, documented and compared behaviors with both lighting systems and found that with the gradual lighting system, all nine macaques were significantly calmer and more relaxed. They all appeared in much better moods, especially in the morning during feeding.

In addition to our behavioral observations, we monitored the animals' cortisol levels throughout the entire study. We used saliva cortisol samples to avoid possible stress reactions associated with collecting blood. The endocrine data mirrored very clearly our behavioral findings. The saliva cortisol concentrations were significantly lower during dawn and dusk phases compared to the same time when the lights were instantly switched on and off. Every one of the nine test animals showed this difference in cortisol concentrations. The results were truly amazing.

Something that has traditionally been completely overlooked in the scientific literature—namely instant versus gradual change from day light to night darkness—can have such a tremendous impact for captive macaques. I imagine myself sleeping peacefully in the dark and all of a sudden bright lights come on and someone says "Time to make the donuts." I'm saying "ugghhh" at that moment versus feeling a warm sunbeam across my cheek and hearing birds twilling in the distance; no bright lights in my eyes, stretching and waking up at my leisure. I get it.

Who is in charge of environmental enrichment?

In your facility, what is the procedure or protocol for adding a new enrichment item? Whom do you ask, who has to grant permission, how long does it take? Who usually brings up ideas for new items—veterinarian, PI, or animal care staff?

Our animal care staff, vet staff, and sometimes the clinical vet propose new environmental enrichment ideas.

Feeding enrichment and food treats for our monks are approved by the vet after we have checked that all fruits and veggies are perfectly safe for the animals.

Devices are also approved by the clinical vet after we have gathered evidence that the object—for example red oak gnawing sticks—is unlikely to cause any harm to the animals. We—the animal care staff—make observations on how practical a particular enrichment device is, how the monks use it and how long it attracts their attention beyond the initial novelty effect.

All approved enrichment items are recorded in a book along with photos.

In our facility we have an environmental enrichment committee that is made up of one veterinarian, our three vet techs, one scientist/investigator from each of our various groups, one toxicologist, one animal care technician and one vivarium floor supervisor. It is a large group of about 14 individuals but we get perspectives from everyone whose work involves the animals. We discuss enrichment devices, foods and changes in regulations for housing and what-not. Any and all new ideas are proposed to the committee and then overall approval comes from the veterinarian. This goes for all species except non-human primates.

We have a separate subcommittee that deals only with enrichment for NHPs. This subcommittee is made up of all individuals who work with the NHPs on a daily basis (about six people); overall approval comes from the vets. The reason we have a separate committee for NHPs is that there always seems to be so much to discuss that our meeting would be too long and those other individuals in our main committee would have no idea what we are talking about since they have nothing to do with the NHPs.

I am trying to get a rough indication of how many primate facilities (and of what size) employ an enrichment tech, or some sort of equivalent position.

I'd love for you guys to help me out and answer the following questions:
1. *How many primates are in your facility (roughly)?*
2. *Do you employ someone with the sole responsibility of providing enrichment?*

If so, is this a technician (or someone else) a full time or part-time position?
3. *If you do not employ someone, how is enrichment handled, and who oversees it all?*

I am working at an academic institution (medical school) that houses a colony of 41 adult macaques. My official title is veterinary technician, but I also serve as the enrichment coordinator. I oversee all aspects of the primate behavior and enrichment program. Some days it's a full-time stint; but most days I squeeze it in with my other duties on campus.

We have about 400 cynos in our facility. The veterinary staff is responsible for providing foraging enrichment and social enrichment (forming new pairs and checking the compatibility of already established pairs).

Everyone plays a role in suggesting novel inanimate enrichment ideas, but the veterinarian has the final say in what we can and cannot use. The husbandry staff ensures that manipulanda are distributed and rotated; the technical staff distributes treats and participates in the various training programs.

Our study directors have input as well, in that some enrichment may interfere with their study goals; if that is the case, they need to submit exemptions to the IACUC for approval.

Our colony currently comprises 50 cynos. We have an NHP technician who is responsible for almost everything that has to do with the cynos, from husbandry to chairing maintenance and environmental enrichment. He is a full-time employee and has additional duties periodically with other species, but focuses most of his time on our NHPs.

Our enrichment program is overseen and approved by vet services.

We have 65 monkeys (cyno and rhesus). Nobody is employed with the sole responsibility of providing enrichment for them. I serve as the behavior manager for all animals at our facility (most of whom are not primates), so I create some of the new enrichment for our monkeys myself. Animal care staff and the research techs take care of most of the daily enrichment—not only for the monkeys but for all our animals.

There are several hundred cynos and a few rhesus in our facility. The administration here just made up a full-time position for me as enrichment coordinator. I mainly take care and supply extra enrichment for monkeys assigned to behavioral studies and for animals with serious behavioral pathologies such as hair-pulling and self-biting. I also work on expanding the environmental enrichment program for the dogs and swine and have started supplying these animals with extra enrichment once a month.

The maximum number of rhesus macaques at our facility is about 80; currently we have 39 animals. We do not have an enrichment tech—though I strongly believe we should. I do most of the enrichment planning. New ideas are approved by the facility management and vet services, and husbandry techs distribute, clean and rotate enrichment gadgets every other week.

We have four full-time enrichment technicians—including me—who are responsible for the enrichment of over 300 New World and Old World primates. It would seem like a sufficient manpower for enrichment, but some of us do not have enough experience when it comes to animals, let alone with the intricacies of enrichment.

A critical specialty that seems to be lacking in many biomedical research labs is an ethologist who is trained to monitor the behavior of the animals. The veterinarian's charge is usually medical/physiological, but there is a void regarding the psychological condition of the animals and how it is affected by specific enrichment strategies.

Our facility has about 3,000 cynomolgus and a few rhesus macaques. There is nobody with the sole responsibility of providing enrichment for these animals. I am employed as behavior technician. I oversee and coordinate the environmental enrichment for the macaques. We rotate between foraging enrichment, regular food treat enrichment, and novel toys/gadgets.

The husbandry staff passes out enrichment items as part of their job most of the time, but some days that is not practical and then I will do it. Hopefully, we'll eventually have an enrichment technician. The only way I get things done is with the assistance of one or two husbandry techs helping me prepare the daily enrichment for the macaques in the afternoon.

There are close to 1,000 rhesus macaques at our facility. We do not have a position specifically designated for enrichment.

Providing inanimate and feeding enrichment is a responsibility that is rotated among the techs on a weekly basis. The difficulty is that at present, it is considered an *if there is time at the end of the day* type of duty, and so only gets done 1–2 times a week. Environmental enrichment is not a priority, so there is hardly any oversight.

Technician time/commitment seems to be the limiting factor to developing and really implementing environmental enrichment for our animals; an enrichment tech would be the simplest, most effective way to accomplish that. As a relatively new doctor at this facility I find it challenging to lobby for the animals. People's minds in this field seem to be strongly anchored to the status quo. Some days I feel I am (slowly) making a difference, other days not at all. If anyone has any specific experience with this type of situation, I would really appreciate any insight or advice.

Every step [even a small one] counts; it is my own experience in this field that with non-judgmental patience and unwavering commitment to your own vision, big changes can be achieved. Sure, you are confronted with obstacles and you have to be ready to take risks at times, but earnestly holding on to your vision, you are bound to reach your goal.

I agree; it is a long road and we must stick to it. Every animal whose status we improve today equals many in the future. If we are not there, who is?!

Environmental enrichment and data variability

Based on your experience with the animals in your charge, would you be concerned that *species-appropriate enrichment is likely to increase data variability, thereby jeopardizing the scientific validity of research data collected from such animals?*

I am of the belief—as most of us probably are!—that depriving animals of species-appropriate enrichment actually makes the data LESS valid. Animals who are bored and frustrated are not good models for research studies. It is not easy to convince some researchers of that, because many want everything to be sterile and unchanging. These conditions are impossible to fulfill when you work with living creatures. Luckily, times are slowly changing and some researchers are becoming more aware of this fact.

I think that environmental enrichment is an essential component of any good animal care program. If species-appropriate enrichment is considered as important as species-appropriate food and professional cleaning, then there should be no issue of variability, as ALL of the animals will be receiving the same kind of environmental enhancement on a consistent basis. In my experience, the problem with variability usually stems from inconsistent enrichment, poorly planned or poorly devised enrichment, and/or a lack of administrative support for the enrichment program.

When we provided protected social contact to our male rabbits, we actually contributed to normalizing their circadian rhythms; the whole rack had the same circadian pattern of activity. I can only imagine that this would remove the variable of free-cycling rhythms we saw in the socially isolated rabbits [Lofgren et al., 2010]. This serves as an example of how enrichment can actually remove variables created by species-inadequate housing conditions.

[Nevalainen et al. (2007) compared changes in growth and selected serum chemistry parameters in pair-housed versus single-housed female rabbits. "No differences in mean profiles were detected; however, weight and APHOS (serum alkaline phosphatase) variances were significantly lower in pair-housed animals." Obviously, environmental enrichment can decrease data variability in certain cases.]

Animal welfare and good science

Who can share experiences or findings that support the following quote from an AALAS meeting abstract? "Good [animal] welfare is good business and good science" (Gaskill et al., 2011).

We had a group of 70 single-caged rabbits on a long-term cholesterol study dragging on for several years. When the animals started developing serious foot problems as a result of sitting on the wire bottoms for such a long time, another employee and I developed a simple floor-housing system and transferred the rabbits in compatible pairs to these much more comfortable quarters. Living in these refined housing conditions improved the animals' inflamed feet very quickly, resolved their chronic pain, and made them friendlier and easier to handle.

That's a great example which deserves to be published. Those responsible for the welfare of animals [investigators and animal care staff alike] in research labs need to be informed about practical Refinement options.

Pigs used to be injected at our facility by simply lifting them off the floor by their hind legs and inserting the needle. Obviously, this was very stressful to the pigs and very disturbing for me while holding them. I knew there must be a better way to work with these intelligent creatures; so I got to thinking.

I developed a simple conditioning program—based on mutual trust and scratching a pig at her favorite spot—that allowed me to give the pigs injections without any special ado and without eliciting noticeable stress reactions. It takes time and patience, but it pays off in better animal welfare and better science (stress-free research data). Happy animals equal good research; I don't think that can be questioned in any way!

I had a very similar experience when I witnessed for the first time the conditions under which blood was drawn from macaques. The animals were forcefully restrained either manually on tables or mechanically in squeeze cages. It was so evident that most of the monkeys suffered extreme anxiety prior to, and intense fear during this common procedure. The literature confirmed that the traditional blood collection procedure triggers significant physiological stress reactions (Elvidge et al., 1976; Bush et al., 1977; Fuller et al., 1984; Hayashi & Moberg, 1987; Landi & Kissinger, 1994); not a good baseline condition to obtain clean, i.e., unbiased research data from the research subject!

As an ethologist and veterinarian I could not go along with this traditional practice. So I developed a safe training technique that allows one person to collect blood from single- and pair-housed rhesus and stump-tailed macaques who cooperate, rather than resist, during this procedure in their familiar home cages.

To make the new Refinement technique more palatable for traditional researchers—who typically don't want to change the way animals have been treated and handled in the past—I conducted supplemental endocrinological studies, which demonstrated that blood samples collected from cooperative macaques do not show the significant increase in cortisol that occurs in animals who are forcefully restrained for this procedure. I also timed myself when training naïve animals. It took me on average 60 minutes to train adult male rhesus or adult female stump-tailed macaques to present a leg for venipunture and show no resistance during subsequent blood collection (Reinhardt, 1991; Reinhardt & Cowley, 1992). This time investment is not unrealistically high, especially when considering the fact that taking a sample from a cooperative animal in her or his home cage takes only a few minutes, plus only one person is needed to accomplish the procedure. In addition, personnel safety is assured because a cooperative animal, unlike a forcefully restrained one, trusts you and hence has no reason to resort to self-defensive scratching or biting.

Yes, I do believe that we can make a difference in terms of animal welfare but also in terms of scientific methodology; they go hand-in-hand.

I worked with seven single-caged adult rhesus macaques who engaged in serious self-injurious biting. These animals were too distressed to be assigned to any research project. I paired all of them successfully with another adult (six pairs) or with an infant (one pair); cumulative time investment was less than one hour per pair formation.

Living with another compatible social companion cured all seven subjects from SIB [good animal welfare]; they turned into normal, i.e., truly social rhesus macaques who could now be assigned to research projects yielding more reliable scientific data [good science].

Are scientific benefits balanced against costs to research subjects?

For most animal technicians and animal caregivers it is very important that the animals in their charge are not subjected to undue pain, stress, distress and suffering associated with and resulting from experimental procedures.

Do you find it helpful listening to in-house presentations of principal investigators/ researchers so that you can get a good picture of the potential scientific benefits and the possible costs for the research subject(s) of an upcoming or ongoing invasive study being conducted with animals in your daily charge?

Our researchers are required to make presentations about projects before they can begin the studies. These informative presentations are mandatory for our staff to attend. I find them very helpful.

Yes, these presentations are very important to get a good picture of what is being investigated. By hearing what the PIs are looking at, you can also advise them—for example, what multiple chairings, extended fasts, and other potentially distressing procedures may do to their model and how these variables can adversely affect their study. When the presentation is open for discussion you can then make suggestions—based on your own experiences—on possible refinements that could buffer or even avoid data-biasing stress reactions of the research subjects.

Unfortunately, I do not always find the time to attend these presentations.

I wish we had those types of informative in-house presentations and discussions. Research protocols are available to us to read, but there is no time for us to review them thoroughly so that we can get an idea of their implications for the research subject's safety and well-being.

It has been my experience that animal caregivers are typically overloaded with work. Yes, they are invited to attend seminars that could inform them about planned and ongoing research in their units, but they are not given the time to do this.

At my facility, it is not that we are not invited, but we don't have these presentations at all. Once in a while a researcher will come to speak to us as part of our continuing education. These talks are given at noon, so we have to take our lunch time to attend.

As an animal technician—not a caregiver—I do not have the time built into my schedule to attend lab seminars. Very few of my colleagues who are actively involved in a study are given the time to attend these lectures. I barely have time to eat lunch some days!

I am also a tech. I have to make time to attend the programs I am responsible for, so that I can offer suggestions to refine studies that would otherwise stress my monks. For the most part, these important sessions are not built into my work schedule.

When you are convinced that a particular study does not have enough scientific merit to warrant that animals experience pain and/or suffer—e.g., the study is redundant, repetitive, poorly designed, inflicts avoidable pain/distress—what do you do?

I am in charge of providing enrichment for the animals assigned to research studies. I've always felt confident concerning the scientific validity of the studies at my facility.

I have never refused a study, because I haven't been in a position where I've had to. However, one of my Ph.D. projects was on enrichment with nursery-reared infant rhesus macaques. I was going to do the study on infants who were already in the nursery for other studies. The expected number of infants wasn't available, so the nursery offered to pull more infants from their moms solely for my study. I refused this option. My goal is to create better lives for the animals that already have less than ideal situations. I do not want to create problems for these animals. So even though my sample size was less than originally planned, I felt good about my decision.

Many, many moons ago, when I was a third-year vet student, we were supposed to participate in a pharmacology practical class. This class involved placing mice on hot plates to assess their reactions to pain with and without analgesics. A group of students—including me—decided that this treatment of mice was unjustified. Accordingly we refused to take part. This was in the days before discussion about the 3Rs was commonplace, and certainly in a time when students didn't question lecturers!

Each of us was required individually to come in front of a panel of academic staff, including the dean of the faculty, to explain our decision. I remember being told that I could not possibly make an informed decision until I was at least three years past graduation. Those of us who would not take part in that treatment of mice were then given an essay to complete instead. There were no formal repercussions, but at the end of the year an unprecedented number of students (I was one of them) failed the pharmacology exam and were required to repeat it—what a mysterious coincidence!

Thirty years after graduation, I still know I made the right decision. Thankfully, we now have robust animal research legislation that ensures this type of educational practice would not gain approval.

I refused to participate in a research project with rhesus macaques that I simply could not condone for ethical reasons, plus I did not see any potential scientific merit in it. It did not come as a big surprise when my annual work contract was not renewed. I was very sad to leave the animals, yet have never regretted that I did not allow myself to be pushed into doing something that I knew was not only inhumane but also unnecessary.

Naming animals in research laboratories

Our facility does not have an official policy regarding naming the animals under our care. It has been discussed recently whether it is appropriate to name terminal animals (specifically dogs), as some of our staff believe that this creates an inappropriate, too affectionate relationship with these animals.

A facility near ours has a strict no-name policy and I believe some of our labs want us to follow suit. Does your facility have a policy on names? What is the reasoning behind the policy?

Traditionally, at our facility we have names for every animal larger than a guinea pig, except the sheep; our sheep are usually here for no more than two weeks. Dogs, rabbits, pigs and NHPs all get names. Some of our researchers named their rats and mice. There are a few pigeons here who have names as well.

We don't have a no-name policy except for the NHPs on GLP studies—to avoid dual-tracking. These GLP monkeys are only identified by their tattoo numbers, which I think is sad.

Several months ago before I left the primate area, I actually held a departmental naming contest for a new shipment of monkeys. It was fun for people to come up with names, starting with the same letter for each group of animals assigned to different study groups.

Naming an animal is typically associated with emotional attachment. I never like to

MISCELLANEOUS 159

hear of people trying to discourage care staff from getting attached to their charges. Yes, it is sometimes hard to say goodbye, and sometimes you don't want to come to work the day that an animal you have grown fond of is leaving, but the benefits outweigh any possible negative effects in my opinion. If you don't have some emotional connection with, and a sense of responsibility for the individuals you are caring for, then what will motivate you to give them your very best?

    Animals who are giving their lives for scientific research deserve the honor of a name.

We have some dogs named and nearly every monkey has a name. As for me, I'm just as attached to a monkey or dog calling them by their ID number or name; either way, I'm still very attached to the individual animal.

Whenever I worked with large numbers of animals, be it cattle, guinea pigs, chicken, bison or macaques, I always named each individual because I found it easier to reliably remember an animal's name rather than his or her official identification number. When taking handwritten field notes of a group of 126 cattle, the names of individual animals [abbreviated in the records]—for example, Alma—popped up in my mind more reliably and faster than this animal's official ID—for example, 74-251.

    I must admit, as a person I prefer to be called by my given name rather than by a number. That's perhaps another reason why I preferred to refer to the names rather than the ID numbers of the animals I studied and/or cared for.

I have been working for nearly 20 years with a colony of common marmosets. All of our animals have names. I encourage students, caretakers and collaborators to choose a name for every newborn monkey.

    To give names to a marmoset introduces an anthropomorphic element into the colony. Although I am aware of the potential variable of this on the actual collection of behavioral data, the benefits are more than the cost. The benefit is to better identify the individual animal—even as a *little person*—and have a personal concern to assure his or her optimal care.

I've found that naming the animals in my charge helps me with compassion fatigue. Identifying an animal by his or her name makes it a lot easier or more effective to relay information. If I say "number 45 looks distressed" it doesn't have as much impact as saying "Sophie is looking distressed."

When I was asked to take care of the well-being of more than 700 macaques, each one of them got a name in addition to his or her identification number. Referring to names, as opposed to numbers, made it easier for me to quickly and correctly identify individual animals.

160

The former director of the Institute for Laboratory Animal Research shares his experience as it relates to naming monkeys: "I was encouraged not to assign names to the many rhesus monkeys in my charge. I was admonished that the animals are research subjects, not pets. The concern was that having names for the animals might blur this distinction between a research subject and a pet. ... It did not seem possible to remain distant—emotionally isolated—from the animals. In fact, the inevitable closeness that resulted from those intimate interactions was precisely what made us capable of doing what we were asked to do. ... Eventually, we all came to know that F49 was Sam, A12 was Rosie, and Z13 was Curious. ... Such attachments are the results of compassionate people doing their job right" (Wolfle, 2002).

Higher versus lower animals

It is not uncommon that animals are categorized into lower-order versus higher-order species. I wonder, do higher-order animals, e.g., monkeys, deserve more animal welfare concern than lower-order animals, e.g., mice?

Based on your own experience with different animal species, would you say that being subjected to a common procedure such as enforced restraint and subsequent blood collection:

(a) is more distressing for monkeys than for cats?

(b) is more distressing for cats than for mice?

Please elaborate briefly on what kind of observations/data/facts you base your response; simply a belief will not suffice.

This is such a difficult question to answer because differences in distress reactions appear at the level of the individual, even within the same species. I have at home two cats who came from a research project. They have been treated pretty much identically since birth. One of them likes to be picked up, cuddled, cradled, etc. The other, who is very affectionate—but only on his terms—will NOT tolerate even being picked up. As soon as I pick him up (physical restraint), he struggles and wriggles to get free and becomes quite distressed.

In terms of your first question, I'm not entirely sure what you mean with higher-order and lower-order.

To my knowledge, the terms higher- and lower-order species/animals are lacking scientific definitions, but I assume that these terms refer to an animal's or a species' standing in the human-created evolutionary taxonomy, degree of intelligence, neurological development, relatedness to humans, charisma and cuteness. In the special context of biomedical research, humans are probably classified as of the highest order, followed by apes, monkeys, companion animals, rabbits, rodents, birds and, finally, cold-blooded animals.

The ethical dilemma arises when presumed lower species, such as mice, are proposed to replace presumed higher species, such as dogs or monkeys, with the implicit assumption that lower species experience less pain and distress than higher species. Here is a quote of a highly respected resource on

MISCELLANEOUS   161

laboratory animal science: "The creation of transgenic animals is resulting in a shift from the use of higher order species to lower order species. ... An example of the replacement of higher species by lower species is the possibility to develop disease models in mice rather than using dogs or non-human primates" (Buy, 1997).

Do we actually know—rather than believe—that a presumed higher animal, such as a macaque, experiences more fear and anxiety during a potentially distressing procedure (e.g., enforced restraint and subsequent blood collection) than a presumed lower animal, such as a mouse?

Personally, I have always had a problem with the terms *higher* and *lower* animals. I believe that it is our responsibility to reduce pain and distress in ALL species that we work with, not just the ones that may be closer genetic cousins.

I would not treat presumed lower-order versus higher-order animals differently but try to cater to their varied species-typical needs. For example, I assume that all vertebrates can feel pain (as opposed to *suffer* pain); so I would give them all analgesia. Very often, I may not be able to alleviate their mental suffering other than by removing them from the painful situation.

I would not use higher- versus lower-order as serious guides, but would rather try to look at an animal's identifiable needs and try to address those accordingly in ALL vertebrates.

For me, it is very clear that so-called lower-order animals deserve as much animal welfare concern as higher-order animals do. People in general are biased in their perspectives. Sadly enough, some individuals still look at mice and rats and see *only rodents,* and do not look at them much more than that.

Fortunately, there ARE also many other individuals—including you and probably all of us on this forum—who look at rodents and other so-called lower-order animals as amazing creatures who deserve our appreciation and, if needed, our unconditional compassion and care.

If you have direct experience with rodents and with monkeys, would you see it as a step towards a more Humane Experimental Technique in the spirit of Russell & Birch (1959) if we developed a certain disease model for mice or rats in order to Replace presumed higher animals—e.g., primates—with presumed lower animals—e.g., rodents? The experiments done with both categories of animals are likely to inflict pain and distress (fear, anxiety).

This question was answered:
› *Yes* by five members of the forum
› *No* by four members of the forum

Some respondents outlined the rationale behind their answer.

My *yes* is based on the assumption that it is in the spirit of Refinement [avoiding/reducing stress and distress] when using a lower species in a potentially distressing research procedure. We are also better equipped to deal with rodent models, and there is less bio-hazard to compromise good animal care. So let's get primate research eliminated first then work on rodents.

From the standpoint of a human primate, the proposal to get primate research eliminated first then work on rodents is understandable. From a less conceptual, more general standpoint of a sentient creature, the proposal may be the opposite: since the number of creatures subjected to painful, distressing and deadly procedures in research labs is much, much higher when rodents are the subjects versus primates, why not get rodent research eliminated first and then work on non-human primates? Each individual creature counts, so the sum total of pain and distress inflicted in research labs is significantly higher when a large number of rodents are used and killed than when a small number of primates are used and killed.

I would have to say *yes* because it takes far more time to handle, dose and treat a non-human primate than a rat or a mouse.

I have a feeling that this isn't going to go over very well, but here goes: *no*. I say this because I strongly believe that the rodent would suffer just as much anxiety and pain as the non-human primate, if not even more.

With primates, we are compelled by law to keep extensive records of what we do. In the U.S., we are required by law to record every procedure we do with a primate. However, because presumed lower-order animals, such as rodents, are not held under such scrutiny, I find that they are not checked as often, handled as well, or given as much attention as the presumed higher-order animals, such as non-human primates. Also, because primates show in their behavioral and emotional expressions that they are probably suffering in a similar manner as do humans, several researchers with whom I have worked tend to sympathize more with the pain of a macaque than they do with the pain of a rat or mouse. The monkey will look depressed; but if the mouse or rat is found alert and responsive they're deemed okay, regardless if it's simply the fight or flight response that kicks in when these creatures are scared. "Well, I tapped on the cage and they ran away; they look just fine to me" is a typical conclusion by an investigator who has not learned how to correctly read the behavior of his animals.

Over the past two days, two major procedures were done in my facility: Vascular surgery on one non-human primate and invasive flap surgery on six rats. The monkey has been tended to by several individuals, me included. The little guy has been checked nearly obsessively for signs of pain and/or distress; he was timely medicated, and pretty much spoiled. The rats? I haven't seen a researcher make an appearance since the gang was brought back into the facility earlier this afternoon. To the best of my knowledge, I am the only individual who has checked on these animals; and there are no records in the room for me to see. Thus, I am unaware of any complications that may have occurred during the procedure; and I am unable to ensure that these rats were all given their appropriate post-op meds. However, I am thankful that they all look okay right now.

Granted, this scenario is not necessarily a given, and I have worked with more than a few really good rodent researchers. However, on the whole, I have witnessed a great deal more complacency with rodent care than with macaque care.

My answer is also *no* because I believe that a rodent experiences anxiety and fear during a distressing procedure to the same extent as a primate does. If strict, species-appropriate animal welfare regulations were put into place also for rodents, then I would possibly change my answer to *yes*, because rats and mice can breed so much more quickly (have shorter life spans, gestation periods and weaning periods, and more offspring at a time), they require less square footage to provide for an adequate living space, and it is easier to find effective enrichment for them that will enhance their behavioral and emotional well-being. I just think that if "the experiments done with both categories of animals are likely to inflict pain and distress," then both categories of animals deserve the same respect toward their welfare, and neither one is more or less deserving than the other.

I must add a disclaimer here: I have had rats as pets for many years, and got to know their personalities very, very well; so my *no* answer may be slightly biased.

I agree with you; rats and mice should not be excluded from legal animal welfare coverage. [In the U.S. rats, mice and birds are explicitly excluded in the legal definition of the term *animal*, so they are not covered by animal welfare regulations (United States Department of Agriculture, 2002).]

Unfortunately, I am excluded from answering this question because I have no direct experience working with monkeys. Having said that, I sit on several ethical committees. If I were to see a research proposal that stated "To reduce suffering, we will use mice as a replacement species to monkeys," I doubt very much I would approve it on *this* basis alone.

Do biomedical studies typically use more subjects when done with rats or mice than when they are done with non-human primates?

If that's the case, what could be the explanation for this bias towards experimentation on presumed lower-order animals versus presumed higher-order animals?

164

The reasons for this bias are money, space and availability. Primates cost a lot, take a lot of space and are not as readily available as rodents, so a study with 50 mice is practicable while a study with 50 primates is not.

There is no statistical logic and I suspect it is based on economics.

In the biomedical industry the numbers are usually driven by the regulatory agencies. The exact numbers of animals used are determined by statistics, economics and/or the agency itself.

[Zbinden (1985) succinctly cautions that investigators "must realize that their important mission ... does not give them an unconditional license to kill as many animals as they wish and hide behind regulatory requirements, testing guidelines and bureaucratic prescriptions for good laboratory practice."]

Are animals aware of themselves?

Many months ago we discussed the usefulness of mirrors as enrichment for non-human primates and inferred from our observations that monkeys have a sense of "I," of "me." That means they are aware of being the creature who is looking back at them from the mirror.

Based on your observations and experience with animals other than primates, would you argue that self-awareness is not restricted to primates, that other animals such as rats, pigs, goats, sheep, dogs, cats and birds also are aware of themselves?

MISCELLANEOUS 165

Not trying to be contrary, I have been wondering about the mirror test. For one thing the test is not positive for all animals. For another, I've seen some of my pet dogs, cats and rats look in mirrors and either groom themselves or look at other things in the room through the mirror. Maybe they recognize the mirror for what it is and, after their investigation realize there is no threat, so they're no longer worried about it. They probably ignore the fact that they are also in the mirror [because they don't identify with their body's reflection].

When I put a mirror in front of my rats, cats and dogs they all give the impression that they simply perceive the reflection of the mirror as movement at the most, perhaps another conspecific, but not as themselves.

I've seen lots of birds preening in mirrors, but it's hard to say whether they perceive the mirror reflection as a mate or as themselves.

Perhaps I will be throwing myself to the wolves, but I have a different take on self-awareness in animals. I do understand the significance of the mirror test, but I am not surprised that many animals do not pass it. I am not convinced that a mark placed on the body is important to all animals and would therefore elicit a reaction of trying to investigate it. Personally, I believe that all living beings possess a sense of self at some level. The only way that I can adequately describe this is to suggest that any being that has a concept of *other* or *that which is not me* must by all logic have a sense of *self*. In order for the cow, dog, goat, cat, monkey or deer to react appropriately and/or recognize the cues from other beings, conspecifics or objects, a sense of *self* is a prerequisite for making decisions based on *self*. When a cat or dog looks down at their own paws, do they not recognize these parts as theirs? Of course tails are a different matter; they sneak up out of nowhere!?

I very much agree. Being aware of her *self* is the very basis of a cow's decision to move out of the way of an *other* cow who has a higher social rank status. Without self-awareness being part of the mental make-up, how could social animals interact with each other in any biologically meaningful manner and, how could a prey animal be able to distinguish the predator as *other* and run away? Yes, self-recognition is a completely different story. Recognizing your *self* when looking into a mirror simply means that you identify yourself with the body that is looking back at you; you believe that yes, this body is me. Humans strongly identify with *their* bodies, hence are always anxious to protect it from real or imagined danger that could possibly lead to the death of their bodies. I have my doubts that animals have such a problem; it would spare them a lot a suffering. They may just flow with life without being possessed by the idea that they have to take good care of the body they happen to have.

Most animals I observe and work with do NOT give the impression that they identify with injured body parts; they simply respond to being injured in a spontaneous but appropriate manner, thereby initiating, promoting and fostering an optimal healing process. I have seen many badly injured wild and domestic animals and was often amazed about the ease with which the individual animal responded to the accident and how amazingly quickly wound healing occurred. I don't think that an animal feels sorry for himself or herself when the body gets harmed; humans certainly do that, thereby creating avoidable suffering for themselves.

Do animals have a sense of humor?

Do animals engage in unusual, playful activities that make them happy? Do they have a sense of humor?

Certainly my dog's enthusiasm for all forms of water to walk or roll in makes me suggest that he does have a sense of humor; not to mention my annoyance when it is a big smelly swamp puddle on the way back to the car, or even better in the car!

    Pigs can be quite funny, exhibiting behaviors that serve no purpose other than getting the humans to react. My favorite example is from almost 17 years ago. We used to exercise our pigs in the afternoons in the dirty hallway. They would run up and down and greet anyone who passed through the hallway with a big slobbery tug on the person's clothes. They also knew who was always around and usually gave them a respectful rub up. When a newbie came down, the response was more overwhelming and included running at full speed and oinking or grunting at the visitor.

    There was an understanding that you did not bring visitors to the facility after 2 p.m. without an appointment because the pigs would be out—and therefore some feces likely to be in the corridor, not a good image for a guest.

I happened to get a new boss during a group shift; he was a scientist who had never supervised an animal group. I explained the need for an appointment after 2 p.m., which he said he understood. About a month later at 3 p.m. one afternoon I hear two of our ladies hauling down the hallway, grunting gleefully. Then I hear the commotion of several voices. I turn the corner and my new boss in his suit and several suited visitors are standing kind of stuck against the wall with two 125-pound piggies tugging on their suits leaving drool marks, brushing up against them and grunting. The pigs had very happy looks on their faces while the visitors were not so amused. I refrained from laughing and called the ladies off with a treat. Now the visitors in their slobbered, smelly suits walked gingerly down the hallway; I gave them the rest of the tour and they left. My boss never came down again without an appointment. I think those two pigs laughed about that for weeks! They were very amused.

Years ago I had a female rhesus who loved to put a plastic pumpkin—the trick-or-treat container kind—on her head and run around bumping into things in her enclosure. She would do this whenever I cleaned with the hose. She would put it on, run around, and take it off, then do it all over again. Always made me laugh, probably her also!

I have seen monkeys pulling each other's tail in a kind of teasing way. The one who pulls the tail exhibits the typical play face, suggesting that this activity serves no serious function but is an expression of humor.

I absolutely believe that animals have a sense of humor.

Our black lab mix, Frodo, has a fun game that he plays, and I think that it is definitely evidence of his sense of humor. He will take his favorite ball and hide it under his bed. He then pounces on his bed, digs through it to recover the ball. Then he repeats the whole thing over and over. We have watched him do this for upwards of an hour; he jumps and rolls around with the ball, it seems to make him so happy!

My most memorable experience was a while back when I worked with young chimpanzees. We had a play space for the youngsters who were between 8 and 18 months old. One female in particular would often take a blanket and put it over her head—like a little *ghost*. She would then chase the other chimps around; they would run away, screaming and laughing. The little ghost would then suddenly pull that blanket off, and the other chimps would laugh and laugh. It looked like a human game of tag, and they definitely seemed to enjoy it. I am always thankful for the time I had with them; they were amazing.

We have a male rhesus who lives in a top-row cage and will pee on you when you add an enrichment device to the cage below. I don't think it's funny but he probably does. You have to watch out for him; he will kind of causally put his hand in the urine stream directing it straight to you. You will feel little sprinkles on your scrub pants and look up and voilà: he is getting the attention he was looking for.

I would call this do-it-yourself enrichment at no extra cost; pretty smart guy!

Retiring and adopting animals who are no longer needed for research

We are spending a lot of money on the humane retirement of a relative small number of chimpanzees, but very little money on the humane retirement of all the other—well over a million—animals who have been used for biomedical research. The resources necessary for offering life-long retirement in species-adequate quarters to 1 research-released chimpanzee would probably be more than enough for offering life-long retirement in species-adequate quarters to 1,000 research-released rats.

What are the reasons behind our bias for primates versus non-primates when it comes to providing animals a well-deserved retirement, after having spent much of their lives promoting biomedical research without their consent? Even within the order of primates, we

tend to be more willing to offer life-long, much more expensive retirement to chimpanzees than to monkeys; why?

As awful as it sounds, it's probably because it is cheaper to retire the few chimps than the huge numbers of other animals.

With chimps being so closely related to humans—not only genetically but also in their appearance and behavior—people working with them on a daily basis probably get very easily attached to them. Along with this emotional relationship comes the ethical dilemma of euthanizing them after they are no longer needed for research. Personally, I also experience this ethical dilemma very strongly when facing the question of euthanizing other animals who, for many people, are perhaps less charismatic than chimpanzees—such as rodents or rabbits.

I would love to see more retirement options for our research animals, irrespective of their relatedness with the human species. Considering the large number of research-released animals who are facing euthanasia every day, efforts to save at least a few in private homes seem to make no real difference; yet, I think each single animal saved and retired does make a big difference—for that particular animal. I have adopted quite a number of rats and two cats who had been released from research. These animals can retire in a safe and caring environment; they do deserve it. I have also found good homes for several bunnies who are no longer used for research at our facility.

I agree that money needs to be set aside for the retirement of laboratory animals besides chimps.

We use animals in research laboratories to help *us*, not them. It would seem quite normal to me that we humans provide these animals the necessary means for their retirement to safe sanctuaries when they are no longer needed for research. In my opinion, that's the least we can do for them in return for their involuntary service to us.

Retiring animals after non-terminal research has to become an integral component of every research proposal. Why not? The funds requested for a research proposal will then cover the expenses necessary to conduct the actual study plus the expenses necessary to assure proper retirement of the animals who served as subjects for that study.

We have several older monkeys who are really just sitting around here after completion of numerous research assignments. My attempts to get them retired have been unsuccessful so far. It is unfortunate that the people *above* don't see these monks every day, so they don't have a close relationship with them. It is heartbreaking when we have to euthanize a 24-year-old monk with whom we have developed a mutual trust relationship over the years, because the animal is now considered useless for research, but occupies a cage that could be used for a younger monkey who can then be assigned to research. I wish we had management that could really take a stand for our monks and make certain that they will be retired for the remainder of their lives after they have served biomedical research endeavors. We owe this to these animals.

My opinion is that if an animal has done service for human health by being used in lab experiments, and is healthy enough to be retired, then she or he should be retired at a safe, professionally caring sanctuary for the rest of his or her natural life. There are some good sanctuaries and more need to be created for all species. Ideally, a combination of federal and private funds should create these sanctuaries so that animals who are no longer needed in research can retire.

The number of animals who survive their last experiment and hence could spend the remainder of their lives retired at sanctuaries is staggering; I would guess there would be well over 1,000,000 rodents, rabbits, dogs, cats, monkeys and other species each year that would have to be processed by an agency and then transferred to sanctuaries. The sheer numbers would make such an endeavor almost impossible, not to speak of the monetary expenses involved in it.

It's true, numbers have important practical implications. But should we not also bear in mind that suffering is an individual experience? When many small animals such as rodents are killed in research laboratories, because it would be burdensome to grant them a retirement, the total amount of suffering inflicted on individual creatures is larger than when a few big animals such as apes are killed in research labs. We easily lose sight of the individual creature when dealing with large numbers of them; but each one of them DOES count because each one of them can suffer. Remember the story of the man throwing the starfish back into the sea?

I am quoting here the famous Starfish Poem for those of you who haven't yet read it: An old man was walking down the beach just before dawn. In the distance he saw a young man picking up stranded starfish and throwing them back into the sea. As the old man approached the young man, he asked; "Why do you spend so much energy doing what seems to be a waste of time?" The young man explained that the stranded starfish would die if left in the morning sun. "But there must be thousands of beaches and millions of starfish," exclaimed the old man. "How can your efforts make any difference?" The young man looked down at the small starfish in his hand and as he threw it to safety in the sea, he said: "It makes a difference to this one!"

We know that we cannot save them all. To give one a chance at life gives one a chance at life. Most of us will see one as better than zero when it comes to saving lives. Just because there were not enough lifeboats on the Titanic did not mean that all should die —you *save* those you can.

It is undeniable that retirement is positive both for the animals and for animal care staff members. The other aspect that is sometimes forgotten in the case of adoptable research animals is the difference these animals can make in the lives of the families that adopt them. The animals go on to become beloved family members, and many of the adopted lab retirees have become certified therapy dogs who visit nursing homes, schools, and hospitals where their mere presence provides comfort to children, the infirm, the elderly and the disabled.

If only more resources were devoted to retirement for animals who could be adopted after they have served biomedical research! Our resources are limited, so until there is more support, the Starfish Poem has to suffice.

I have the starfish story posted on my locker as inspiration for my everyday work.

Throwing starfish back to safety in the sea is exactly what individual institutions can do with adoption policies for animals who are no longer used for biomedical research. We won't be adopting out all of them, but would be making a big difference to at least some of them, and these all count individually.

Yes, that's a very important point: when we are making sincere efforts to have animals— *even* the little and perhaps less charismatic ones—adopted/retired after termination of their research involvement rather than kill them for convenient reasons, we are making a difference not only to the individual animals but to ourselves and our staff as well. I remember the shift in mood when animal care staff had to kill some of their rabbits or a group of mice. For some people, this final procedure can be very hard, regardless of the animal's size or species. It is difficult for any sensitive and reasonable person to accept that healthy animals be killed after they have served the biomedical industry, rather than given away for adoption to suitable homes or retired for the rest of their lives in professional sanctuaries.

Retiring or rehoming rodents is quite possible for small labs—I've done it. I wish there would come a time when we would provide a safe retirement for all animals; until then we can make sincere efforts to retire at least some of them, and this is already practicable.

I know it's a suggestive question, but do researchers who are making use of animals to promote their scientific career not have a basic ethical responsibility to make sure that the subjects who provided them with their valuable research data are granted a humane and safe retirement after they have served their research endeavor? Personally, I find it ethically unacceptable when research institutions first make use of animals and then kill them—monkeys, rodents, rabbits, cats, dogs and cold-blooded animals alike—when they are no longer funded under a research protocol. Would it not be fair if a portion of the requested funds for a research project with animals had to be allocated upfront for the research subjects' life-long retirement? The number of animals used would certainly drop dramatically, but I think this would not automatically lead to a breakdown of biomedical science; the opposite could be true: scientists would have to improve their scientific research methodology to a point where they truly have to use only the minimal number of animals to obtain reliable data and statistically sound results.

Does your institution have an active adoption program for animals who are no longer assigned to funded research protocols? Did you ever rescue/adopt an animal or several animals who were scheduled to be killed after termination of a study? Was it difficult to get permission from your facility to do this? How did you adjust to living with one of your former charges as a pet in your home?

In 1981 I was a student at FIOCRUZ's [*Fundação Oswaldo Cruz*] first Technician in Parasite Biology program. We had classes

in which we learned technical skills using rabbits; one of those rabbits could not be used and hence was to be killed; he was a young animal. I asked to save him and got permission to take the rabbit home with me.

I named him Bingo. He seemed to be content in his new home and there was no room that he did not explore. Bingo had his cage where he would go to drink water and get his food. He usually slept on a piece of cloth on the floor, although sometimes he slept inside his cage. I bought some rubber toys for him that he enjoyed playing with. He loved carrots more than anything else. Bingo developed an affectionate relationship with us and he would often come to the door and kind of greet us like a dog when we arrived at home after work. He died when he was about nine years old.

We have an adoption program at our facility and routinely adopt out rabbits, and occasionally mice, rats, ferrets and frogs. The ferrets and rabbits are always spayed or neutered before they move out, and we offer spay/neuter for rats and mice if the owner is interested. So it was very easy for me when

I adopted a female NZW rabbit a number of years ago. She adjusted very well to her living environment; she was a little shy at first but soon came out of her shell and enjoyed the space and freedom that the small cage in the laboratory had not offered. It was great to watch her just being a quasi-free rabbit! I had never owned a rabbit before, and it wasn't until I had her that I realized how truly inadequate the standards are for rabbit caging in the research lab.

We have a beagle and a cat who were both adopted from a research facility at which I used to work. Both were babies when I brought them home, so they adjusted very well. They are a constant source of happiness for both me and my husband. When I worked at this facility, we had a fairly active adoption program, and from what I understand it is still going strong.

I think it is high time that retirement and adoption become valid post-research endpoints. Simply because these animals were bred for research does not mean that their lives should end when their protocol is done; besides, what better way to say "thank you for your service to mankind!"

We had a strict no-adoption policy here but then we had an accidental beagle pregnancy. There was a lot of back and forth about the dam's future and if she would be allowed to go to term. Thankfully, she was too far along and too many staff members knew about the case—I may have helped with that. We quickly developed an adoption program and had the IACUC and a legal team draft the necessary protocol and contracts. We were then able to adopt the mom out and found a good home for her. We also got enough people lined up to take all of the puppies.

I was there when the mom went into labor and was able to see all the puppies being born! The runt had to be bottle fed by the third day; mom just could not supply enough milk. I decided to become the mom for one of them and brought her home; our little Belle. In fact, my son hooked up my computer to our TV last night and found the birthing video I took, so it's very fitting that this discussion came up today. Belle has been a great addition to the family!

One good thing I got from my previous job is the joy that my three beagles Gabby, Dotty and Scrunchy bring me every day. I had named them at work, even though it was very much frowned upon. When they were at risk of being culled, I successfully pleaded with

the management of the research facility to let me adopt them.

When I brought these three girls home, they were about 1½ years old. They have adapted perfectly well to living outside of the research lab. They seem to be content and happy, and this makes me also happy, very happy.

I have brought home all kinds of animals in the past, from rodents and rabbits to livestock; my yard is full of chickens from my ocular research days. I will say it seems to be much easier to implement adoption programs in the academic setting versus contract research.

It's perhaps not as rare as one might think that researchers, like you, get so attached to their animals that they adopt them rather than have them killed after the termination of the study. When searching for photos for this book, I came across this photo with the following caption:

Susie and Hazel Snuggle
Susie is an 8-yr-old pig who spent her first six months used in research studying lung ventilation. The researcher fell in love with her and didn't want to see her slaughtered. Hazel, Susie's companion, is a 6-yr-old pig rescued from a cruelty case where she was malnourished and infested with mange mites.

I can share two photos of one of my adopted rats.

This cute little female ended up not being used on study, so she was slated for euthanasia. We have an adoption procedure here, so fortunately for both of us, I was able to bring her home. Her name is Tulip; she is a hairless rat. The veterinarian checked her over and deemed her ready to *go home*.

I live in a drafty old farm house in Pennsylvania. Of course, Tulip was used to the wonderful climate-controlled atmosphere of her old home, and bringing her to my house was quite a change. One evening, while watching TV with her inside my sweater, I decided it was time for her to have a sweater (or two) of her own. While she watched, I knitted a sweater out of some leftovers I had stashed away. It turned out very nice but wasn't quite the right style and color for her.

This week I went on a search for something that was more her style and found some lovely soft pink sock yarn. With the new yarn, I crocheted her something more to our liking. The second photo is the final creation that Tulip is modeling. As you can see, she is very happy wearing this nice warm sweater that everyone refers to as her "tutu."

When I had finished my studies of a marmoset breeding colony I wanted to start a sanctuary and bring home the retired breeders, but I was able to work out arrangements with a local zoo instead. A few times a year, I go to visit these animals who have served my research endeavors.

We had a researcher who planned to continue a study with two cats who had lived at the facility already for one year. After no research was conducted with these animals in the course of the next five years, I approached the administration asking if I could take them home. My boss at the time convinced the researcher to let me adopt these cats, who were already nine years old at that time.

They are currently 15 years old and living in the lap of luxury! They love lying in the sunshine, begging for food—not sure where they learned to do that!—and harassing my other two cats. Although they didn't die, they have definitely gone to heaven!

I am reading between the lines that your decision to adopt these two cats not only made the two cats happy but probably made you even happier. When we are kind to others, be it animals or people, we unknowingly are kind to ourselves; a very simple, basic equation for happiness.

176

Human-animal trust relationship

It is my experience that I must first have established a mutual trust-based relationship with an animal before I can safely and successfully train him or her to cooperate with me during procedures. I wonder, once an animal has learned that she or he can trust you, will the animal also trust other humans?

I would like to believe that this is true, but I don't think it is across the board. The rabbits with whom I worked on a long-term study were always nervous and fearful when they saw a lab coat, even though they had learned to trust me; they would follow me around and sit on my lap and allow me to gently hold them. I never wore the ordinary white lab coat but a surgery or isolation gown. In hindsight I can see that I should have worn a lab coat now and then; this perhaps could have desensitized my rabbits to the dreaded lab coats.

Like you, I've seen animals who were completely comfortable with care staff but get very stressed by the presence of a lab coat.

Lab coats would warrant a separate discussion. Many animals simply learn through fear-inducing, often life-threatening experiences that humans wearing lab coats are potential predators; how could they trust me when I approach them the first time, wearing a lab coat even though I have good intentions?! Unfortunately, the lab coat quickly becomes an acute alarm signal for so many animals confined in research labs. I have learned that very early in my career and always refused to wear the professional coat when working with animals. When I trained non-human primates to cooperate during traditionally distressing procedures I never wore the dreaded lab coat but a dark blue or brown coverall. This simple adjustment in my attire made it very, very easy to quickly gain the animals' trust, i.e., the foundation of subsequently training them to work with rather than against me.

By the way, the white coat has a similarly alarming effect for human patients as it has for animals in laboratories.

I also wear scrubs when working with our monkeys but over that I put on the same blue jacket as the care staff. The care staff can do amazing things with these animals, they have a very strong bond with them. I believe that trusting their care staff can help them to develop trust also in other people—but never completely! I still find with my monks that I have to earn the remarkable trust they have in their caregivers.

I recently rescued a little shaggy dog (five years old) from a puppy mill, where she was routinely kicked. She exhibited the typical tail-between-the-legs, cowering in a corner, and growling when I got her. Obviously, she was not very happy with humans.

I took her straight to the vet and then to the groomer to get the scary stuff done before I took her home.

Now de-fleaed, de-wormed, vaccinated and groomed—which had nothing to do with me, in her eyes—I carried her around all day and worked with her intensively. I let her sleep in my bed—between me and my husband—on her own pillow.

Within only a few days she has bonded with me tightly and doesn't leave my side. Happy-go-lucky when she's with me, completely house-trained, no more growling but rather wagging her little tail high in the air; she can't seem to do enough to please me—until someone else approaches her; then she cowers and growls.

We've had her now for two weeks and I have been trying to socialize her with volunteers AND my husband. So far not much luck. She now tolerates my husband lying next to her in HER bed, but that's about it, even though he truly has put a lot of effort into making her feel at ease with him.

I've experienced similar outcomes with others animals, a couple of horses, a few dogs, and monkeys included, where I earned their trust and was able to harmoniously work with them, but even after several years no one else could ever seem to gain their trust. But I've also had animals who did learn to trust other people after I had earned their trust; in my own experience, however, such cases are pretty rare.

It has been my experience with adult rhesus and stump-tailed macaques who I have trained to cooperate during procedures that the animals subsequently cooperated in the same manner also with other familiar care personnel and even with strangers, provided these were dressed in a coverall similar to mine and approached the animals with friendly intentions. I was not in the room when these other people interacted with the animals, which suggests that the animals also truly trusted them.

I think a person's intention is the key to gaining an animal's trust. Animals pick up our unspoken intention very precisely; there is no cheating! But yes, wearing a lab coat may provoke so much conditioned fear in an animal that he/she can no longer sense our genuinely friendly intentions and resorts to aggressive self-defense.

Based on your own experience, would you discourage the establishment of mutual trust-based human-animal relationships in the research lab because scientific data collected from the animals could be influenced by such affectionate relationships? Could it be that an animal who is treated like a standardized research object yields statistically more reliable data than an animal who is treated like a sensitive research partner?

A trust relationship with the animals in my charge is very likely to have an effect on the research data collected from them; the effect will be in the spirit of Refinement, so I would certainly not discourage but, on the contrary, I would strongly encourage friendly interactions with animals who are assigned to research. There is no published evidence and, in fact, there is no good reason to believe that a mutual trust-based relationship between human caregiver/handler and animal can affect research data in any negative way. It will help buffer stress and distress reactions to being handled, and as such will make research data collected from the animal more reliable, i.e., less affected by the uncontrolled extraneous variable of stress.

I do believe that the human-animal relationship affects the quality of scientific research outcomes. This relationship can either be positive, when the animal trusts the human or negative, when the animal has fear of the human. Research data collected from an animal who trusts the human handler are likely to be free of data-biasing stress or distress reactions, while data collected from a fearful animal will be compromised by uncontrolled physiological stress reactions. Establishing a trusting relationship with animals assigned to research is, therefore, a refinement of research methodology; it helps to minimize or eliminate stress as a data-biasing variable.

Establishing a trust-based relationship with my cynos and rhesus monkeys also provides valuable enrichment, not only for the animals but also for me. I find it relatively easy to work with monkeys who trust me; they are more cooperative during procedures, which means they are much less—in many cases, not at all—stressed during sample collection, and hence yield scientifically more reliable data.

Our cynos and rhesus respond differently to humans they feel comfortable with, as opposed to humans they either don't know or don't like. With humans they trust, they are happier, more relaxed, and easier to work with. Data collected from them are bound to be different than data collected from animals who are upset, angry or frightened. The question as to whether these data are scientifically more valid is, in my opinion, redundant, and I say "If the monkey ain't happy, ain't NOBODY happy."

Dealing with emotional fatigue

I am seeking input, feedback and ideas about how to prevent or mitigate compassion fatigue and burnout when working with animals in research laboratories. What strategies or activities might offer support for the more difficult, emotional aspects of that work—for example, euthanizing animals as part of your job or witnessing your animals suffering distress during certain research procedures?

Some really good people leave science because they simply cannot continue killing animals. Many of these individuals are the type of kind, sensitive, empathetic people we want to be the ones working with the animals. Of course, some people who really do not want to do this work anymore should just not do it.

Perhaps some of you are willing to share what do you do personally to deal with compassion fatigue, and what approach do you take with your staff to address this issue?

I am new to this field. It has been quite a challenge to adjust to my new work environment; there were, and still are, many days of tears. In my position as animal behaviorist I am doing all I can to make the animals' lives better, even if it hurts me in the end. Their lives have purpose and while they are here, it is my job to see that they are happy.

But how do you avoid an eventual burnout?

This is something that I continue to struggle with, despite being in the industry for 16 years. Although, I think that if I ever get used to it, and no longer struggle, it will absolutely be time for me to find a new path.

For me it is also very difficult emotionally to work in the lab on studies where I know the animals are going to be put down at the end of it all. I was told once not to get attached and that there will always be another animal to replace the one who was sacrificed for an important cause. As much as this might be true, it has never been easy for me to accept it. There were many animals I would get close to and when they went down I cried. Unfortunately, it doesn't ever get easier with time; you just learn how to deal with the given situation in your own way. Fortunately, other technicians at my facility are very understanding and sympathetic. We are sharing these sad experiences together.

How we cope with the unavoidable, emotionally very disturbing situations of our daily work is a personal thing, but talking about it certainly helps, knowing you are not alone. It has been my experience that it is beneficial to find colleagues who feel as I do, so we can share those feelings and find out about coping mechanisms that perhaps we never thought of. It is through forums like this that those connections can be made.

I benefit from exercise, meditation, eating healthy, getting plenty of sleep and of course a lot of thinking about the difference I make with the care of my animals in the here and now, and how I can improve their lot in the future. I hope some day animals will no longer be used in biomedical research.

I know we have all experienced compassion fatigue; it's almost unavoidable when you have a deep love for the critters you are working with but have little or no control over their fate.

Compassion fatigue is something that must be dealt with on an individual level, as it is a form of distress. Over the years I have found a few things that help me to deal with this reoccurring situation. First, if it stops hurting then it's time to get out. I admit that after more than a decade, I still find myself shedding tears each time I put down a rat. And honestly, earlier this week, as I euthanized a rabbit to whom I had given critical study-care for more than two weeks, my heart broke and more than one tear fell. But, I collected myself and moved forward, as the research staff was waiting on me. I mention this because I have found that to allow myself to grieve is the best way for me to cope. I always say "goodbye" and "thank you" and give a final scratch to any animal prior to administering the final anesthetic shot for euthanasia preparation. In the end, I know what I do is for a purpose, and believing in that purpose is what gets me through. And, if at the end, I know I did the best for the animal that I could, and gave her or him the best

life possible under the circumstances given, it helps me to deal with the final moments better.

Prior to euthanizing any animal, I speak with the caretakers, the researchers, the other vet staff and sometimes even the cage wash staff so they are aware of what will happen, and give them time to grieve the loss they may feel. I never remove an animal from a room for euthanasia (rodents included) without notifying the care staff. To unexpectedly find an empty cage or empty pen can be shocking and heartbreaking. We spontaneously create bonds with these animals; to lose one or several of them can hurt just as much as losing a family pet. I allow staff to get mad and speak their minds about the situation regardless of how negative the diatribe may become. I allow this in order for them to cope; it's a kind of release of the tension resulting from extreme frustration.

Occasionally, when I can see it's all starting to get to people, I hold "group." I put out a notice, say clearly that it's voluntary, and allow anyone who wants to come to share how they feel about a certain situation. I've had as little as one person show up, and during especially hard times—such as several weeks of terminal dog work—I've had a turn-out of 15 animal care employees. I think it's good to help oneself and others get these emotional issues off the chest; it is my experience that it certainly makes the rest of our daily routines smoother.

Emotional fatigue is a very important topic for all of us, and I agree that you should get out if it stops hurting. Finding ways to cope is an individual task, but knowing there is a group of others (supportive others!) going through similar experiences is a great relief. I love that you notify staff before euthanasia—nothing worse than going into a room and finding a buddy gone, not having had the chance to say goodbye. I try to convince myself that I won't get *that* close to another monkey, but then I fall in love all over again.

It can be very tough.

It was my dream to go to vet school; a car accident, leaving me with years of chronic pain, changed that route for me. I went into lab animal research, as I was an animal science major. The constant euthanizing of rats and mice was very depressing for me, so I switched to biomedical research with monkeys. You may think that it is harder to euthanize a monkey; for sure it is, however, we very rarely euthanize, and that is a great relief for me. When we do have to put a monkey down, I am usually not involved with the procedure. The vet knows how much I love these animals and keeps me out of it; it's very hard for me. I go visit with my other monkeys; they cheer me up and I know they need me to keep them happy.

I always go to say goodbye to the monkey who is scheduled to be euthanized the next day and give her or him extra treats and goodies, but for sure I end up in tears; it's not easy to deal with euthanasia when the animal is not sick or suffering in any way!

The continued challenge and the affectionate bonds I have with my monkeys keep me going; I truly love these animals and want to make their lives better.

As veterinary staff we usually know in advance when a terminal procedure is in the works and we can in most cases give the care staff a "free" day. This gives them time to prepare; but also, they can spoil the monkey (within reason) during that day. This is a good time to celebrate the animal's contribution and thank them for their service and sacrifice. As for me, each time one of my guys completes the journey I say, I will never become attached again, but all I have to do is see those big brown eyes and I fall again. I think the key for me is to know I am giving them the best I have.

Communication between animal care staff and investigators

It is my experience that many—unfortunately not all—animal caretakers and animal technicians are sincerely concerned about the well-being of the animals in their charge. They are typically in a more qualified position than the principal investigators to assess environmental and procedural factors that can cause avoidable stress and/or distress for research-assigned animals. The PIs very often have little or no direct contact with the animals of their research program, so they don't really know if uncontrolled extraneous variables are confounding the data obtained from them, thereby necessitating a relatively large number of research subjects to achieve statistical significance of the research results.

As animal caretaker or animal technician, do you communicate with the PI about your ideas of avoiding or at least minimizing certain housing- and handling-related variables of which the investigator is not aware?

We as animal techs have very little contact with the PIs. This can sometimes be frustrating, especially when our views regarding animal welfare and species-appropriate housing and handling are kind of ignored by the PI who, typically, is not aware of his research subjects' fears, discomfort and distress.

[Herzog (2002) reported in the *Institute for Laboratory Animal Research Journal* that "I have spoken with some animal care staff who have complained about investigators who rarely set foot in their institution's animal colony and who appear to regard research animals as organ repositories. In addition, some researchers show little understanding of the ethical problems faced by technicians."]

Over the years I have realized that the varying ways people get into the research field and how the scientific research staff is reviewed and rewarded (grants and pay raises) make many PIs unaware of the fact that the welfare of their animal subjects has a significant impact on the scientific validity of research data obtained from them. Animals are not a focus or even an interest of most investigators.

A prestigious researcher concedes in the journal *Laboratory Animal Science* that "investigators think only briefly about the care and handling of their animals and clearly have not made it an important consideration in their work" (Traystman, 1987).

Having worked as a scientist with many scientists in the course of more than 30 years, I must say that the genuine scientific motivation of researchers is very often clouded by a fierce career-oriented competition that leaves little room for so-called sentimentalities such as compassion for animals and making sure that animals are properly housed and carefully handled during procedures.

I can discuss animal welfare concerns openly with several researchers without fear of them getting defensive or reading more into what I am saying than necessary. There are other researchers who are immediately defensive and can quickly become combative. They will take it as a personal offense when your suggestion implies that they may not be aware of everything. I'm sure there is a government program that requires research institutions to hire a certain quota of these difficult-to-like people.

Regardless of how PIs react, I always bring up concerns that I or my staff have regarding the welfare of animals on study; I feel it is my ethical obligation.

I may not speak on behalf of all animal care techs, but it seems to me that we have chosen our profession because we are fascinated by animals and have compassion for them. Researchers on the other hand chose their work with animals not because they have feelings for them but because animals can help them promote their professional careers in biomedical science. To find a consensus between these rather opposite goals and achieve a harmonious cooperation between both sides is, indeed, a big challenge.

I was actually hired specifically to deal with this challenge. The lack of communication between the animal-house staff and the researchers in our lab was slowing down research projects and husbandry procedures. The lab hired me to act as a go-between, to develop and implement a more constructive communication process.

I am responsible for managing all ongoing research. This implies that I inform myself thoroughly about each research project, its scientific goals and its implications both for the research subjects—the animals—and for the attending care personnel. I spend 30–40% of my time working directly with the animals

and co-coordinating with the administration projects that investigators wish to conduct. The animal care staff discusses problems and ideas with me, and we can usually address any issues quickly and effectively as a team. My boss is a pretty typical PI in that he doesn't really have any contact with the animals and not much with animal care staff. He tells me what he wants to achieve with the research project and then gives me free reign to work out with the animal staff the most effective way of implementing it.

When my boss talks about a research project he starts talking about apoptosis proteins and blocking biochemical pathways and rescuing phenotypes by altering something on a molecular level; I can understand this language because I trained myself. When the animal care staff talk about their work with the animals they are usually thinking more about environmental enrichment, housing, general health status and animal handling. I am also working with animals, so I can understand their language, as well, and facilitate a constructive communication and respectful understanding between both parties.

I make an effort to get to know all the research staff involved in a project and inform myself properly about the project's methodology and goal and its animal welfare implications. I am respectful of the investigators' preferences and needs while never losing track of my goal, which is the health and well-being of the animals. This really doesn't have to be an *us* versus *them* mentality—or maybe I'm just too doggone friendly.

Overall, our PIs are as reasonable as the protocols allow. I know that I am always listened to if there is an animal welfare issue. My observations are taken seriously, and I have a free hand to do what I think is best for the animals within the given confines of the research. To give an example: I asked one of our investigators recently if I could rearrange the cage locations of his monkeys to get a submissive and very nervous monkey out of direct eye contact with the most dominant monkey in the room. His exact words were "whatever you think can make her more comfortable please, arrange it."

I start out relationships with researchers and their staff by telling them that they have the right to do whatever experiments they are approved for, and I am here to help them in any way I can, but also that I will be looking after the health and well-being of their animals and will not hesitate reporting off-protocol incidences when I see them. Most researchers and their staff understand this and accept it, some accept it without really understanding the practical implications; nevertheless, all researchers do respect the animal care techs and work with them together as a team in order to get high quality research data from animals who are healthy and receive the best possible care.

It gives me great pleasure to say that most of my institution's investigators are very open to animal welfare issues. In particular, I have an investigator who is amazingly concerned that his animals—dogs and swine—receive the best possible care. He's a physician by trade and is one of the most attentive and caring researchers I have ever worked with.

He has actually come to me to inquire as to how certain animals should be handled in order to keep them *happy* prior to even submitting a new protocol. He ensures that his entire staff is well acquainted with the critters they are working with and how to treat and handle them, so that they experience the least possible stress during procedures. He is also open to having his animals adopted after they have been released from research. During the past three years, seven of his dogs and one swine found loving homes. Finally, he allows me to train his animals so that they don't have to be subjected to avoidable restraint, especially during long-term procedures such as timed medication administration and blood draws. He's actually willing to pay the extra per diem to have the animals brought to our facility to give everybody sufficient time to make the animals feel at ease and cooperative during handling procedures. He is really remarkable.

Thank you for sharing this exquisite example of a truly responsible and caring investigator. I do think that it requires some humbleness on the part of a PI to seriously listen to animal caretakers and technicians who, after all, have so much more first-hand experience with animals before, during and after research is done with them.

I have always been lucky to work very closely with PIs. I am very concerned about the behavioral health and the emotional well-being of animals in research, and we occasionally have our differences but are always able to come to a meeting point that is agreeable to both of us. I think the biggest issue is trust. If the investigators can trust that I have a good understanding of their research project and that I am supportive of its goal, chances are that they will not only listen to my concerns but also respect them. Yes, some homework has to be done to get a good background understanding of the PI's research; only then can I expect that the PI will try to understand my position and possible animal welfare concerns. In order to be taken seriously, I have to have valid points and be prepared to back them up with published data or my own observations. I study the IACUC protocols and look for ways that procedures can be refined to benefit the animals without compromising the study before the study even starts. I have suggested many changes that have solved serious problems with study design; it is my experience that PIs appreciate that. Now they often ask for my input when planning new research projects.

The important thing is not to see each other as different departments but as cooperating team members representing different, equally valid positions but with the same goal.

Because I've worked in nearly every aspect of the facility, I have run across a host of different levels of interactions and personalities. Husbandry staff is hired to provide basic care for the animals. They need to work on a schedule created by others, stay within a timeline as dictated by a time clock or supervisor, and do very repetitious assignments in order to keep their positions. Research staff is hired to conduct scientific projects; PIs need to write grants, publish papers and attend scientific meetings in order to keep their position and advance

their career. It can be hard for the two groups to get on the same wavelength. And the techs, well, we have to float between husbandry staff and research staff, who at times seem to speak completely different languages and are unable—or unwilling—to respect each other's viewpoints and goals.

I tend to have a very good relationship with the PIs, their graduate students and staff. However, there are a few PIs who seem to believe that their academic education makes them superior. I have run into investigators who treated me as if I have *lower intelligence* simply due to the fact that I wear a blue coat and punch a clock. To communicate with such individuals is really a challenge, sometimes impossible.

Professional satisfaction

Assuming you did not choose your profession by accident, what motivated you to become an animal technician, animal caregiver or clinical veterinarian at a biomedical research lab?

Before you started that career you probably had specific expectations for being satisfied and happy with your work. When you look back, were/are these expectations fulfilled? If you had a choice, would you choose the same professional career again?

I have loved being around animals ever since I can remember. When I was four years old I had the dream to become a veterinarian. In sixth grade I wanted to work with marine animals; I have always been fascinated by whales, dolphins, sharks and other big fishes, but I did not know how to turn this fascination into a profession that would allow me to make a living.

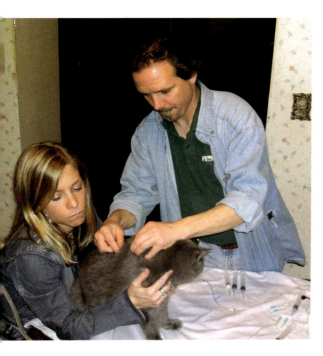

I became a vet tech and worked several years at a veterinary hospital; it was a great experience for me. I also had thoughts of becoming a zoo keeper, which probably led me to my present work with monkeys who essentially are wild animals—just out of the zoo setting. I do love my work with the monks, however I dislike the politics of a large pharma company. I try to stay focused on my job and my animals and very much enjoy working with the vets, but every now and then upper management decisions and requests make me sad and frustrated.

I am very glad that I can work for the well-being of the animals I care for; that's what keeps me going.

This past Sunday I went swimming in a shark tank and in a stingray pool; watched the sharks swim right by a few inches away from my face. I LOVED it; they are some of the most amazing, beautiful creatures in the ocean. The rays were equally awesome; they were begging like puppies to be fed and petted. After that experience I wished I had pursued my dream of working for an aquarium. But at the same time, I can say that I also very much enjoy making a difference for the monks I care for.

We are cut from the same cloth. I too wanted to go into marine biology, but living landlocked and without the means to get myself to a coast, I gave it up. Now I am working, like you, with monkeys!

I have swum with stingrays as well, and it was one of the most magical things I have ever done! They really do seek out physical contact with the humans that are near. I went to the Georgia Aquarium this past fall, and finally got to see mantas up close; swimming with them is on my wish list before I depart this Earth.

I knew since I was a very young girl that I wanted to work with animals. I would tell people that I want to be a *vegetarian*, because I couldn't pronounce the word *veterinarian*.

As I got older, I worked at our local SPCA [Society for the Prevention of Cruelty to Animals] and realized that I prefer working directly with the animals rather than taking the responsibility of a vet.

During vet tech college years, I needed to do two placements: one month in a vet clinic and one month in a biomedical research lab. There was a technician at my research placement who had gone to my college and graduated the year ahead of me. She worked part-time at my placement as well as at another facility to which she took me one day and showed me around. On the last day of my research placement she was offered a full-time position, therefore her other job became available, and I got the job!

Before I became a tech, I knew that I wanted to work with animals in laboratories and help them in whatever way I could, whether by making them feel better, enriching their lives, or finding them a home when they are no longer needed for research. In my current position I definitely do that, providing the monks in my care the best and most entertaining living conditions. I hope they are all happy.

My dream is to open up a retirement center for primates who have been used in research.

I was desperate to work with animals and had originally intended to make a living in agriculture, never having heard of the term *animal technician*, let alone being aware of animals in research. Couldn't get a career in agriculture—and it had to be a career, not just a job. In those days girls milked cows and fed calves or poultry; this was not my cup of tea. When I applied for a job with our Ministry of Agriculture, I received the recommendation that I should work in the laboratories and, more by luck, was placed in a lab that did all the veterinary vaccine testing; and the rest, as they say, is history.

It's been a fantastic 40+ years. I worked with most species, apart from reptiles. I've been frustrated sometimes, but this was always related to the limitations that were imposed on me because of being technical, female or working class. I had an excellent first boss who taught me that anything was in reach if I worked hard for it: all in all not a bad lesson for life.

Working with animals has never disappointed me and I've never stopped learning from them—and yes, I would choose the same kind of work again.

When I was getting ready to graduate, our instructors set up many tours of private veterinary practices large and small, and a few research facilities nearby, helping us to find our career path and make contacts. I swore to myself NEVER, after touring the research facilities, not because I saw anything awful, I just thought working with animals in a research lab wasn't for me. My instructors tried to help me see differently, assuring me that I would probably find far more cruelty and suffering out in veterinary practice than in the biomedical research industry. Well, I was young and stubborn and went into a largely exotic veterinary practice; and my instructor was right, I did see much cruelty and

suffering. After several years the veterinary practice restructured, and I quit under duress.

With a mortgage and no job, I found myself turning to the university to find a job and finally entered the research animal world. I became the primary caregiver for a large group of long-term rabbits. I learned a lot about myself, and where I stand in caring for and loving those girls. I still remember each and every one. I did end up going back to private practice for a bit, but quickly came to the realization that, while I function well in that capacity, caring for the lives of research animals is truly where my satisfaction lies. Medicine only goes as far as nature will allow, but the care one can give for the research animal lies in your own hand—with facility budget a limiting factor at times. Presently I am caregiver for macaques.

I am anxiously waiting for the time when animal research ceases to exist. I wish for it, but in the meantime I come home every night with a sense of fulfillment. Knowing that I make a big difference for the animals in my care; showing them affection and respect is an honor and one of the greatest satisfactions in my professional career. Until biomedical research moves on to non-animal models, I'm staying.

References

Aidara D, Tahiri-Zagret C and Robyn C 1981 Serum prolactin concentrations in mangabey (*Cercocebus atys lunulatus*) and patas (*Erythrocebus patas*) monkeys in response to stress, ketamine, TRH, sulpiride and levodopa. *Journal of Reproduction and Fertility 62*: 165–172

Akiyama K and Sutoo D 2011 Effect of different frequencies of music on blood pressure regulation in spontaneously hypertensive rats. *Neuroscience Letters 487*(1): 58–60

Anderson JR and Chamove AS 1984 Allowing captive primates to forage. *Standards in Laboratory Animal Management. Proceedings of a Symposium* pp. 253–256. The Universities Federation for Animal Welfare: Potters Bar, UK

Andrews K, Morelli N, Ruesterholz E, McAllister S and Coleman K 2012 The use of bedding for groups of rhesus macaques. *35th Meeting of the American Society of Primatologists Scientific Program*: Abstract # 138 https://www.asp.org/meetings/abstractDisplay.cfm?abstractID=4391&confEventID=4502

Archard GA 2012 Effect of enrichment on the behaviour and growth of juvenile *Xenopus laevis*. *Applied Animal Behaviour Science 39*: 264–270

Baumans, V, Coke, CS, Green, J, Moreau, E, Morton, D, Patterson-Kane, E, Reinhardt, A, Reinhardt, V and Van Loo, P (eds) 2007 *Making Lives Easier for Animals in Research Labs: Discussions by the Laboratory Animal Refinement & Enrichment Forum*. Animal Welfare Institute: Washington, DC http://www.awionline.org/pubs/LAREF/LAREF-bk.html

Bazille PG, Walden SD, Koniar BL and Gunther R 2001 Commercial cotton nesting material as a predisposing factor for conjunctivitis in athymic nude mice. *Lab Animal 30*(5): 40–42

Bentson KL, Capitanio JP and Mendoza SP 2003 Cortisol responses to immobilization with Telazol and ketamine in baboons (*Papio cynocephalus/anubis*) rhesus macaques (*Macaca mulatta*). *Journal of Medical Primatology 32*: 148–160

Blom HJM, van Tintelen G, Baumans V, van den Broeck J and Beynen AC 1995 Development and application of a preference test system to evaluate housing conditions for laboratory rats. *Applied Animal Behaviour Science 44*: 279–290

Boccia ML 1989 Preliminary report on the use of a natural foraging task to reduce aggression and stereotypies in socially housed pigtail macaques. *Laboratory Primate Newsletter 28*(1): 3–4 http://www.brown.edu/Research/Primate/lpn28-1. html#maria

Brent L and Weaver D 1996 The physiological and behavioral effects of radio music on singly housed baboons. *Journal of Medical Primatology 25*: 370-374 http://www.awionline.org/lab_animals/biblio/ jmp25-3.htm

Brooks DL, Huls W, Leamon C, Thomson J, Parker J and Twomey S 1993 Cage enrichment for female New Zealand White rabbits. *Lab Animal 22*(5): 30–38

Brown MJ and Nixon RM 2004 Enrichment for a captive environment - The Xenopus laevis. *Animal Technology and Welfare 3*: 87–95

Bryant CE, Rupniak NMJ and Iversen SD 1988 Effects of different environmental enrichment devices on cage stereotypies and autoaggression in captive cynomolgus monkeys. *Journal of Medical Primatology 17*: 257–269 http://www.awionline.org/lab_animals/biblio/ jmp17-2.htm

Bush M, Custer R, Smeller J and Bush LM 1977 Physiologic measures of nonhuman primates during physical restraint and chemical immobilization. *Journal of the American Veterinary Medical Association 171*: 866–869

Buy M 1997 *Transgenic Animals, (From CCAC Resource Supplement, Spring/Summer 1997. Used with permission of CCAC).* http://people.ucalgary.ca/~browder/transgenic.html

Chamove AS and Anderson JR 1979 Woodchip litter in macaque group. *Animal Technology 30*: 69–74 http://www.awionline.org/lab_animals/biblio/at-cham.htm

Chopra PK, Seth PK and Seth S 1992 Behavioural profile of free-ranging rhesus monkeys. *Primate Report 32*: 75–105

Chu L, Garner JP and Mench JA 2002 Pair-housing rabbits in standard laboratory cages: The relative importance of social enrichment. *Contemporary Topics in Laboratory Animal Science 41*(4): 114

Cozens M 2006 Evaluation of the provision of hay to guinea pigs at GlaxoSmithKline. *Animal Technology and Welfare 5*: 31–32

Crockett C. M. and Gough GM 2002 Onset of aggressive toy biting by a laboratory baboon coincides with cessation of self-injurious behavior. *American Journal of Primatology 57*: 39 http://www.asp.org/asp2002/abstractDisplay.cfm? abstractID=306&confEventID=285

Crockett CM, Shimoji M and Bowden DM 2000 Behavior, appetite, and urinary cortisol responses by adult female pigtailed macaques to cage size, cage level, room change, and ketamine sedation. *American Journal of Primatology 52*: 63–80

da Costa AP, Leigh AE, Man MS and Kendrick KM 2004 Face pictures reduce behavioural, autonomic, endocrine and neural indices of stress and fear in sheep. *Proceedings of the Royal Society, Series B 271*: 2077–2084 http://rspb.royalsocietypublishing.org/ content/271/1552/2077.full.pdf+html

DiVincenti L, Rehrig A and Wyatt J 2012 Interspecies pair housing of macaques in a research facility. *Laboratory Animals 46*: 1170–1172

Donnelly MJ, Wickham A, Kulick A, Rogers I, Stribling S, Strack A, Doerning B and Feeney W 2007 A refinement of oral dosing in the common marmoset (*Callithrix jacchus*). *American Association for Laboratory Animal Science [AALAS] Meeting Official Program*: 45

Elvidge H, Challis JRG, Robinson JS, Roper C and Thorburn GD 1976 Influence of handling and sedation on plasma cortisol in rhesus monkeys (*Macaca mulatta*). *Journal of Endocrinology 70*: 325–326

Fante F, Baldan N, De Benedictis GM, Boldrin M, Furian L, Sgarabotto D, Ravarotto L, Besenzon F, Ramon D and Cozzi E 2012 Refinement of a macaque transplantation model: application of a subcutaneous port as a means for long-term enteral drug administration and nutritional supplementation. *Laboratory Animals 46*: 114–121

Fuller A 2009 Successful pair housing of female New Zealand White rabbits. *Tech Talk 14*(4): 2–3 http://www.aalas.org/pdfUtility. aspx?pdf=TT/14_4.pdf

Fuller GB, Hobson WC, Reyes FI, Winter JSD and Faiman C 1984 Influence of restraint and ketamine anesthesia on adrenal steroids, progesterone, and gonadotropins in rhesus monkeys. *Proceedings of the Society for Experimental Biology and Medicine 175*: 487–490

Gaskill B, Winnicker CL, Garner JP and Pritchett-Corning K 2011 The naked truth: Breeding performance of outbred and inbred strains of nude mice with and without nesting material. *American Association for Laboratory Animal Science [AALAS] Meeting Official Program*: 150–151

Gerold S, Huisinga E, Iglauer F, Kurzawa A, Morankic A and Reimers S 1997 Influence of feeding hay on the alopecia of breeding guinea pigs. *Zentralblatt für Veterinärmedizin 44*: 341–348

Glenn AS and Watson J 2007 Novel nonhuman primate puzzle feeder reduces food wastage and provides environmental enrichment. *American Association for Laboratory Animal Science [AALAS] Meeting Official Program*: 45 http://nationalmeeting.aalas.org/pdf/2007-abstracts.pdf

Guhad FA and Hau J 1996 Salivary IgA as a marker of social stress in rats. *Neuroscience Letters 27*: 137–140

Hall CS and Ballachey EL 1932 A study of the rat's behavior in a field: A contribution to method in comparative psychology. *University of California Publications in Psychology 6*: 1–12

Harr J, Coyne L, Chaudhry A and Halliwell RF 2008 A study of the impact of environmental enrichment on *Xenopus Leavis* oocytes. *AWI Quarterly 57*(3): 25 http://www.awionline.org/awi-quarterly/2008-summer/study-impact-environmental-enrichment-xenopus-laevis-oocytes

Hayashi KT and Moberg GP 1987 Influence of acute stress and the adrenal axis on regulation of LH and testosterone in the male rhesus monkey (*Macaca mulatta*). *American Journal of Primatology 12*: 263–273

Hedge TA, Saunders KE and Ross CA 2002 Innovative housing and environmental enrichment for bullfrogs (*Rana catesbiana*). *Contemporary Topics in Laboratory Animal Science 41*(4): 120–121

Hennessy MB 1997 Hypothalamic-pituitary-adrenal response to brief social separation. *Neuroscience and Biobehavioral Reviews 21*: 11–29

Herzog H 2002 Ethical aspects of relationships between humans and research animals. *ILAR [Institute for Laboratory Animal Research] Journal* 43(1): 27–32 http://dels-old.nas.edu/ilar_n/ilarjournal/43_1/Ethical.shtml

Hilken G, Willmann F, Dimigen J and Iglauer F 1994 Preference of Xenopus leavis for different housing conditions. *Scandinavian Journal of Laboratory Animal Science 21*: 71–80

Hu Y, Xu L, Yang F and Fang P 2007 The effects of enrichment with music or colorful light on the welfare of restrained mice. *Laboratory Animal and Comparative Medicine 2*: 71–76

Huang-Brown KM and Guhad FA 2002 Chocolate, an effective means of oral drug delivery in rats. *Lab Animal 31*(10): 34–36

Kilcullen-Steiner C and Mitchell A 2001 Quiet those barking dogs. *American Association for Laboratory Animal Science [AALAS] Meeting Official Program*: 103

Kirchner J, Hackbarth H, Stelzer HD and Tsai P-P 2012 Preferences of group-housed female mice regarding structure of softwood bedding. *Laboratory Animals 46*: 95–100

Krohn TC, Salling B and Kornerup Hansen A 2011 How do rats respond to playing radio in the animal facility? *Laboratory Animals 45*: 141–144

Lambeth SP, Perlman JE and Schapiro SJ 2000 Positive reinforcement training paired with videotape exposure decreases training time investment for a complicated task in female chimpanzees. *American Journal of Primatology 51*(Supplement): 79–80

Line SW, Clarke AS, Markowitz H and Ellman G 1990 Responses of female rhesus macaques to an environmental enrichment apparatus. *Laboratory Animals 24*: 213–220

Line SW and Morgan KN 1991 The effects of two novel objects on the behaviour of singly caged adult rhesus macaques. *Laboratory Animal Science 41*: 365–369

Line SW, Markowitz H, Morgan KN and Strong S 1991 Effect of cage size and environmental enrichment on behavioral and physiological responses of rhesus macaques to the stress of daily events. In: Novak MA and Petto AJ (eds) *Through the Looking Glass. Issues of Psychological Well-being in Captive Nonhuman Primates* pp. 160–179. American Psychological Association: Washington DC

Lofgren JL, Wrong C, Hayward A, Karas AZ, Morales S, Quintana P, Vargas A and Fox JG 2010 Innovative social rabbit housing. *American Association for Laboratory Animal Science [AALAS] Meeting Official Program*: 131

Luchins KR, Baker KC, Gilbert MH, Blanchard JL and Bohm RP 2012 Manzanita wood: A sanitizable enrichment option for nonhuman primates. *Journal of the American Association for Laboratory Animal Science 50*(6): 884–887

Lynch R 1998 Successful pair-housing of male macaques (*Macaca fascicularis*). *Laboratory Primate Newsletter 37*(1): 4–5 http://www.brown.edu/Research/Primate/lpn37-1.html#pair

Marr JM, Gnam EC, Calhoun J and Mader JT 1993 A non-stressful alternative to gastric gavage for oral administration of antibiotics in rabbits. *Lab Animal 22*(2): 47–49

McDermott J and Hauser MD 2007 Nonhuman primates prefer slow tempos but dislike music overall. *Cognition 104*: 654–668

McLean CB and Swanson LE 2004 Reducing stress in individually housed sheep. *American Association for Laboratory Animal Science [AALAS] Meeting Official Program*: 144

McMillan JL, Maier A and Coleman K 2004 Pair housing adult female rhesus macaques: Is it always the best option? *Folia Primatologica 75*(Supplement): 395

Mori Y, Franklin PH, Petersen B, Enderle N, Congdon WC, Baker B and Meyer S 2006 Effect of ketamine on cardiovascular parameters and body temperature in cynomolgus monkeys. *American Association for Laboratory Animal Science [AALAS] Meeting Official Program*: 178

Nevalainen TO, Nevalainen JI, Guhad FA and Lang CM 2007 Pair housing of rabbits reduces variances in growth rates and serum alkaline phosphatase levels. *Laboratory Animals 41*(4): 432–440

Núñez MJ, Mañá P, Liñares D, Riveiro MP, Balboa J, Suárez-Quintanilla J, Maracchi M, Méndez MR, López JM and Freire-Garabal M 2002 Music, immunity and cancer. *Life Sciences 71*: 1047–1057

Ogura T and Tanaka M 2008 Preferred contents of movies as an enrichment method for Japanese macaques. *Primate Eye 96*: 99

Parrott RF, Houpt KA and Misson BH 1988 Modification of the responses of sheep to isolation stress by the use of mirror panels. *Applied Animal Behaviour Science 19*: 331–338

Reinhardt V 1991 Training adult male rhesus monkeys to actively cooperate during in-homecage venipuncture. *Animal Technology 42*: 11–17 http://www.awionline.org/lab_animals/biblio/at11.htm

Reinhardt V and Cowley D 1992 In-homecage blood collection from conscious stump-tailed macaques. *Animal Welfare 1*: 249–255 http://www.awionline.org/lab_animals/biblio/aw1blood.htm

Reinhardt V 1993a Enticing nonhuman primates to forage for their standard biscuit ration. *Zoo Biology 12*: 307–312 http://www.awionline.org/lab_animals/biblio/zb12-30.htm

Reinhardt V 1993b Promoting increased foraging behaviour in caged stumptailed macaques. *Folia Primatologica 61*: 47–51

Reinhardt V 1993c Using the mesh ceiling as a food puzzle to encourage foraging behaviour in caged rhesus macaques (*Macaca mulatta*). *Animal Welfare 2*: 165–172 http://www.awionline.org/lab_animals/biblio/aw3mesh.htm

Reinhardt V 1994 Pair-housing rather than single-housing for laboratory rhesus macaques. *Journal of Medical Primatology 23*: 426–431 http://www.awionline.org/lab_animals/biblio/jmp23.htm

Rourke C and Pemberton DJ 2007 Investigation of a novel refined oral dosing method. *Animal Technology and Welfare 6*(1): 15–17

Russell, WMS and Burch, RL 1959 *The Principles of Humane Experimental Technique*. Methuen & Co.: London, UK http://altweb.jhsph.edu/pubs/books/humane_exp/het-toc

Scales M and McDonald KM 2011 Factors influencing the preferred nesting location of laboratory mice. *American Association for Laboratory Animal Science [AALAS] Meeting - Abstracts of Poster Sessions*: 31

Seier JV, Mdhluli M, Collop T, Davids A and Laubscher R 2008 Voluntary consumption of substances of unknown palatability by vervet monkeys: a refinement. *Journal of Medical Primatology 37*: 88–92

Sherwin CM, Haug E, Terkelsen V and Vadgama M 2004 Studies on the motivation for burrowing by laboratory mice. *Applied Animal Behaviour Science 88*: 343–358

Shyu WC, Nightingale CH, Tsuji A and Quintiliani R 1987 Effect of stress on pharmocokinetics of amikacin and ticarcillin. *Journal of Pharmaceutical Sciences 76*: 265–266

Streett JW and Jonas AM 1982 Differential effects of chemical and physical restraint on carbohydrate tolerance testing in nonhuman primates. *Laboratory Animal Science 32*: 263–266

Sutoo D and Akiyama K 2004 Music improves dopaminergic neurotransmission: demonstration based on the effect of music on blood pressure regulation. *Brain Research 1016*: 255–262

Torreilles SL and Green SL 2007 Refuge cover decreases the incidence of bite wounds in laboratory south african clawed frogs (Xenopus laevis). *Journal of the American Association for Laboratory Animal Science 46*(5): 33–36

Traystman RJ 1987 ACUC, who needs it? The investigator's viewpoint. *Laboratory Animal Science 37*(Special issue): 108–110

Tse MM, Chan MF and Benzie IF 2005 The effect of music therapy on postoperative pain, heart rate, systolic blood pressures and analgesic use following nasal surgery. *Journal of Pain and Palliative Care Pharmacotherapy 19*(3): 21–29

United States Department of Agriculture 1995 *Code of Federal Regulations, Title 9, Chapter 1, Subchapter A - Animal Welfare*. U.S. Government Printing Office: Washington, DC http://www.aphis.usda.gov/animal_welfare/downloads/awr/awr.pdf

United States Department of Agriculture 2002 *Animal Welfare Regulations Revised as of January 1, 2002*. U.S. Government Printing Office: Washington, DC http://www.access.gpo.gov/nara/cfr/waisidx_04/9cfrv1_04.html

United States Department of Agriculture 2005 *Code of Federal Regulations, Title 9, Chapter 1, Subchapter A - Animal Welfare*. U.S. Government Printing Office: Washington, DC http://www.access.gpo.gov/nara/cfr/waisidx_05/9cfrv1_05.html

Van Loo PLP, Blom HJM, Meijer MK and Baumans V 2005 Assessment of the use of two commercially available environmental enrichments by laboratory mice by preference testing. *Laboratory Animals 39*: 58–67

Walters SL, Torres-Urbano CJ, Chichester L and Rose RE 2012 The impact of huts on physiological stress: a refinement in post-transport housing of male guineapigs (*Cavia porcellus*). *Laboratory Animals 46*: 220–224

Watson SL, Shively CA, Kaplan JR and Line SW 1998 Effects of chronic social separation on cardiovascular disease risk factors in female cynomolgus monkeys. *Atherosclerosis 137*: 259–266

Wells DL, Graham L and Hepper PG 2002 The influence of auditory stimulation on the behaviour of dogs housed in a rescue shelter. *Animal Welfare 11*: 385–393

Wheatley, BP 1999 *The Sacred Monkeys of Bali.* Waveland Press: Prospect Heights, IL

Whitney RA and Wickings EJ 1987 Macaques and other old world simians. In: Poole TB (ed) *The UFAW Handbook on the Care and Management of Laboratory Animals, Sixth Edition* pp. 599–627. Churchill Livingstone: New York, NY

Wolfle TL 2002 Introduction. *ILAR Journal 43*(1): 1–3
http://ilarjournal.oxfordjournals.org/content/43/1/1.full

Zbinden G 1985 Ethical consideration in toxicology. *Food and Chemical Toxicology 23*: 137–138

Zhou K, Li X, Yan H, Dang S and Wang D 2011 Effects of music therapy on depression and duration of hospital stay of breast cancer patients after radical mastectomy. *Chinese Medical Journal 124*: 2321–2327

Photo Credits

page 1
top: Conner Downey | Flickr Creative Commons
middle: Chess | Flickr Creative Commons
bottom: Maia C. | Flickr Creative Commons

page 2
top: Jon Ross | Flickr Creative Commons
middle: cskk | Flickr Creative Commons
bottom: Julie German | Flickr Creative Commons

page 3
top, left: jiva | Flickr Creative Commons
top, right: Jennifer Lamb | Flickr Creative Commons
bottom: SV Johnson | Flickr Creative Commons

page 4
top: Don Burkett | Flickr Creative Commons
bottom: James & Mary Bilancini | Flickr Creative Commons

page 5
left: Chris Connolly | Flickr Creative Commons
right: Dan Buczynski | Flickr Creative Commons

page 6
Gunnar Sigurður Zoega Guðmundsson | Flickr Creative Commons

page 7
top: Dominique Bergeron | Flickr Creative Commons
bottom: Kris Miller | Flickr Creative Commons

page 8
Adam Gerard | Flickr Creative Commons

page 9
Patricia van Casteren | Flickr Creative Commons

page 10
top: Mark & Andrea Busse | Flickr Creative Commons
bottom: John Sibley | Flickr Creative Commons

page 11
top: PINKÉ | Flickr Creative Commons
middle: Christian | Flickr Creative Commons
bottom: Betty B | Flickr Creative Commons

page 12
top: Melanie Cook | Flickr Creative Commons
bottom: Si Beedie | Flickr Creative Commons

page 13
top: Memily | Flickr Creative Commons
bottom: Understanding Animal Research | Flickr Creative Commons

page 14
Alan Porter | Flickr Creative Commons

page 15
top and middle: Mike Suarez
bottom: Jinjian Liang | Flickr Creative Commons

page 16
top: Quinn Dombrowski | Flickr Creative Commons
bottom: Angela N. | Flickr Creative Commons

page 17
Marilyn Johnson

page 18
Marji Beach | Flickr Creative Commons

page 19
Matt McCants | Flickr Creative Commons

page 20
top: Charles Roffey | Flickr Creative Commons
bottom: Daniel Lloyd | Flickr Creative Commons

page 21
Ben Cooper | Flickr Creative Commons

page 22
Adam Fagen | Flickr Creative Commons

page 24
Cowgirl Jules | Flickr Creative Commons

page 25
top: @abrunvoll | Flickr Creative Commons
middle: hello-julie | Flickr Creative Commons
bottom: John Donges | Flickr Creative Commons

page 26
left: Louis DiVincenti
right: Cory

page 27
left: Mike | Flickr Creative Commons
right: Sarah Macmillan | Flickr Creative Commons

page 28
top: Andrew T. Sullivan | Flickr Creative Commons
bottom: Jo Christian Oterhals | Flickr Creative Commons

page 29
top: Kairon Gnothi | Flickr Creative Commons
bottom: Frédéric Salein | Flickr Creative Commons

page 30
Kristen Ankiewicz | Flickr Creative Commons

page 31
top: Novartis | Flickr Creative Commons
middle: Amy Selleck | Flickr Creative Commons
bottom: Rebecca Lai | Flickr Creative Commons

page 32
top: Amanda Majakoski | Flickr Creative Commons
middle: Dean Thorpe | Flickr Creative Commons
bottom: Lindsay | Flickr Creative Commons

page 33
top: rattyfied | Flickr Creative Commons
bottom: Ian Crowther

page 34
left: Dario Linsky | Flickr Creative Commons
right: bclinesmith | Flickr Creative Commons

page 35
left: Grinwik | Flickr Creative Commons
right: Anonymous

page 36
top, left: Halsted Bernard | Flickr Creative Commons
top, right: Pehpsii Altemark | Flickr Creative Commons
bottom: Tamara Godbey

page 37
left: StevelnLeighton | Flickr Creative Commons
right: Evelyn Skoumbourdis

page 38
Maureen Hargaden, Hoffmann-La Roche Inc.,
Nutley, NJ; ©Roche

page 39
Understanding Animal Research | Flickr Creative Commons

page 41
Brittany Randolph | Flickr Creative Commons

page 42
left: Thomas Huston | Flickr Creative Commons
right: Frankenstoen | Flickr Creative Commons

page 45
left: Mark Philpott | Flickr Creative Commons
right: Morgan | Flickr Creative Commons

page 47
Understanding Animal Research | Flickr Creative
Commons

page 48
Derek Lyons | Flickr Creative Commons

page 49
top: Rachel | Flickr Creative Commons
middle: Maureen Hargaden, Hoffmann-La Roche
Inc., Nutley, NJ; ©Roche
bottom: Martin | Flickr Creative Commons

page 50
left: Zebra Pares | Flickr Creative Commons
right: Mark Ordones | Flickr Creative Commons

page 52
Tasayu Tasnaphun | Flickr Creative Commons

page 53
Jean-Etienne Minh-Duy Poirrier | Flickr Creative
Commons

page 54
Phoenix Dark-Knight | Flickr Creative Commons

page 55
top: Rourke C.
bottom: Radagast | Flickr Creative Commons

page 57
top: Andrew Iverson | Flickr Creative Commons
middle: Kynan Tait | Flickr Creative Commons
bottom: Hans Splinter | Flickr Creative Commons

page 58
ravend | Flickr Creative Commons

page 60
Anonymous

page 61
top: Keith Survell | Flickr Creative Commons
bottom: nmb | Flickr Creative Commons

page 62
Keith Survell | Flickr Creative Commons

page 63
left: Jim Robinson | Flickr Creative Commons
right: Robobobobo | Flickr Creative Commons

page 64
left: André Mouraux | Flickr Creative Commons
right: Franco Rios | Flickr Creative Commons

page 66
Silke | Flickr Creative Commons

page 68
Michele Cunneen

page 69
top: Keith Survell | Flickr Creative Commons
middle: Keith Survell | Flickr Creative Commons
bottom: MPR | Flickr Creative Commons

page 70
Max Maass | Flickr Creative Commons

page 73
top: Geoff Gallice | Flickr Creative Commons
middle: Emmanuel Keller | Flickr Creative
Commons
bottom: Polly Schultz

page 74
Viktor Reinhardt

page 75
Angelika Rehrig

page 76
Richard Lynch

page 77
Polly Schultz

page 80
left: Paul Mullett | Flickr Creative Commons
right: Eric Kilby Flickr | Flickr Creative Commons

page 81
Viktor Reinhardt

page 82
left: Marcie Donnelly
right: Polly Schultz

page 83
Polly Schultz

page 85
JillannRawlins ONPRC

page 86
Hari Prasad Nadig | Flickr Creative Commons

page 87
left: Anne Roberts | Flickr Creative Commons
middle: Paul Wittal | Flickr Creative Commons
right: Jerry Dohnal | Flickr Creative Commons

page 88
Michael Ransburg | Flickr Creative Commons

page 89
top: Maia C | Flickr Creative Commons
middle: Jens Vilhelm Rothe | Flickr Creative Commons
bottom: Tambako The Jaguar | Flickr Creative Commons

page 90
top: velo_city | Flickr Creative Commons
bottom: Niamh Cotter | Flickr Creative Commons

page 91
left: DEMOSH | Flickr Creative Commons
middle: Adam Wise | Flickr Creative Commons
right: Pat Murray | Flickr Creative Commons

page 92
Andrew West | Flickr Creative Commons

page 93
Polly Schultz

page 94
top: Michael Goodine | Flickr Creative Commons
middle: Richard Forward | Flickr Creative Commons
bottom: Mark Bellingham | Flickr Creative Commons

page 95
left: Gordon Anderson | Flickr Creative Commons
right: Wan Taquddin | Flickr Creative Commons

page 96
Polly Schultz

page 97
David Bygott | Flickr Creative Commons

page 99
Dean Thorpe | Flickr Creative Commons

page 101
Anonymous

page 103
Polly Schultz

page 104
Abney Dawn

page 106
Anonymous

page 108
left: Polly Schultz
right: Arno Meintjes | Flickr Creative Commons

page 110
left: Anonymous
right: Polly Schultz

page 111
Jim Kenefick | Flickr Creative Commons

page 112
left: Polly Schultz
right: Alison Kulick

page 113
C.K. Koay | Flickr Creative Commons

page 114
left: Jürgen Seier
right: Rikard Lagerberg | Flickr Creative Commons

page 117
Viktor Reinhardt

page 120
Polly Schultz

page 121
Alison Kulick

page 125
Michael Keen | Flickr Creative Commons

page 127
Viktor Reinhardt

page 131
Bob Dodsworth

page 133
Ross Huggett | Flickr Creative Commons

page 135
Bob Dodsworth

page 135
Viktor Reinhardt

page 140
Dave See | Flickr Creative Commons

page 141
Bob Dodsworth

page 145
Viktor Reinardt

page 148
top: Henry Burrows | Flickr Creative Commons
middle: Carol Browne | Flickr Creative Commons
bottom: Mack Lundy | Flickr Creative Commons

page 149
Anonymous

page 160
Bob Dodsworth

page 162
Peter Kemmer | Flickr Creative Commons

page 164
neofedex | Flickr Creative Commons

page 165
top: Polly Schultz
bottom: Carol Vinzant | Flickr Creative Commons

page 167
Andrew | Flickr Creative Commons

page 168
Jasper Nance | Flickr Creative Commons

page 171
Susan DeBoer

page 172
Caleb Malcom | Flickr Creative Commons

page 173
left: Ben Cooper | Flickr Creative Commons
right: Bubblejewel96 | Flickr Creative Commons

page 174
Ali Moore

page 175
top: Marji Beach | Flickr Creative Commons
bottom: Mary Rambo

page 176
top: Jonathan Crowe | Flickr Creative Commons
bottom: furecatstef | Flickr Creative Commons

page 178
Priscilla | Flickr Creative Commons

page 179
Viktor Reinhardt

page 180
Shavon McKinstry | Flickr Creative Commons

page 182
John Morris | Flickr Creative Commons

page 187
mcsquishee | Flickr Creative Commons

page 188
Lori Horwedel | Flickr Creative Commons

Index

Acclimation to humans, rabbits: 67–71
adopting animals after termination of
    research: 169–177
African green monkeys: 97–98, 113–114, 132–133
Alpha-dri: 41
apples, pigs: 19–22
arboreal dimension, primates: 73–75
bandage, social housing: 147–148
bedding material, mice: 41
Beta-chip: 41
blood collection, pigs: 22–24
blood collection, primates: 125-127, 131
burnout: 180-183
cage change, mice: 40–41, 44–45
cage change, rats: 53–54
capturing caged primates: 120–124
cardboard, rabbits: 57–58, 60–61
cats: 1-3, 148, 173, 177
cold-blooded animals: 29–30
cynomolgus macaques: 74–75, 94–97, 99,
    101–104, 107–108
cynomolgus macaques: 122–124, 129, 149–150, 152
dog feeders: 14
dogs: 3–9, 149, 165, 173–174, 177–178
double-decker cage: 36

elevated resting surface, cats: 1–2
elevated resting surface, dogs: 10–12
elevated resting surface, primates: 73–74
elevated resting surface, rabbits: 60–61
emotional fatigue: 180–182
Enviro-dri: 37
environmental enrichment, administration: 152–155
environmental enrichment, cats: 1–3
environmental enrichment, cold-blooded
    animals: 29–30
environmental enrichment, data variability: 155–157
environmental enrichment, dogs: 3–9
environmental enrichment, goats: 27–28
environmental enrichment, pigs: 16–19
environmental enrichment, rabbits: 57–61, 60
environmental enrichment, rats: 31–33
environmental enrichment, sheep: 25–25
environmental enrichment, usefulness
    of this term: 43
feces eating, macaques: 139
Flagyl: 111
foraging devices, primates, commercial: 77–81, 85
foraging devices, primates, custom-made: 80–82
foraging enrichment, primates: 77–91
foraging enrichment, rodents: 45–46

frogs: 29–30
frozen produce, primates: 89–89
gnawing stick, primates: 91–93
goats: 27–28
group formation, primates: 98–99
guinea pigs: 46–48
hair pulling, macaques: 137–138
hammock, primates: 73
hammock, rats: 32–33
hand feeding, mice: 55
hand feeding, pigs: 19–21
hand feeding, rabbits: 68, 70, 173
hand feeding, rats: 31–32, 35
hay, guinea pigs: 47–48
higher versus lower animals: 161–165
human attention, cats: 1–3
human attention, dogs: 4–8,13
human attention, monkeys: 9, 13–14, 135, 140–144
human attention, pigs: 21
human attention, rabbits: 67–71
human attention, rats: 31–32, 35
humor in animals: 167–169
injection of sedatives, primates: 127–128
investigator, relationship to research animals:
      163, 172, 175, 183–186
lab coats, not in animal rooms: 69–70, 177–178
long-tailed macaques: 74–75, 94–97, 99,
      101–104, 107–108
long-tailed macaques: 122–124, 129, 149–150, 152
lower versus higher animals: 161–165
mangos, primates: 86–87
marmosets: 113, 120–122, 151, 177
marshmallow, marmosets: 81–82, 121
marshmallow, pigs: 21
metronidazole: 111
mice: 35–46, 48–52, 55
mirror, dogs: 165
mirror, pigs: 25–27
mirror, primates: 165
mirror, rabbits: 59
monkey see, monkey do: 118–120
music in animal rooms: 149–151
naming animals: 159–161
nest, mice: 37–43

nest, rabbits: 64–65
nesting location, mice: 44
nesting material, mice: 37–43
Nestlet: 37–39, 42
noise making, rabbits: 58–59
oral dosing, pigs: 21–22
oral dosing, primates: 108–115
oral dosing, rabbits: 71–72
oral dosing, rats: 54–55, 111
pain, signs of pain, rabbits: 72
pair formation, female rabbits: 62–64
pair formation, pigs: 15–16
pair formation, primates: 96–108
pair housing, macaques, census: 105
pair housing, macaques, different species: 107–108
pair housing, macaques, partner separation:
      105–107
paper, dogs: 4
paper, mice: 37, 39, 42
paper, pigs: 17
paper, rabbits: 57–58, 60
paper, rats: 33
perch, primates: 73–74
phased lighting: 151-152
pigs: 5, 15–24, 148, 149, 156, 167–168, 175
play area, dogs: 12–14
play area, pigs: 18
play area, primates: 75–76
play area, rabbits: 59–60
popcorn, primates: 83–84
Prang, pigs: 20, 23–24
Prang, primates: 89, 128
primates: 73–144
principal investigator, relationship to research
animals: 163, 172, 175, 183–186
produce, primates: 86–91
professional satisfaction of animal care
      personnel: 187–189
quarantine time, primates: 135–137
rabbits: 57–72, 149, 155–156, 156, 172–173
radio sounds in animal rooms: 149–151
rats: 31–35, 46, 53–55, 151, 162, 175
retiring animals after termination of research:
      169–177

rhesus macaques: 74–75, 94–95, 101–104,
107–108, 110,
rhesus macaques: 112–113, 117, 122, 127, 129,
131, 135, 149–152
rooting, pigs: 16–19
saliva collection, primates: 128–129
self-awareness in animals: 165–167
self-biting, macaques: 138–139, 157
sheep: 25–27
shelter, frogs: 29
shelter, guinea pigs: 46–48
shelter, mice: 35–43, 48–49
shelter, rabbits: 58
shelter, rats: 32–33, 48–49
Shepherd Shack: 38–39, 42–43, 49
social housing, dogs: 5–6
social housing, female rabbits: 62–64
social housing, male rabbits: 65–67
social housing, rats: 32–33
squirrel monkeys: 98–99, 124–125
starfish poem: 171
stereotypical locomotion: 145–147
teeth brushing, dogs: 7
toys, cats: 2
toys, dogs: 3–6
touching non-human primates: 141–144
training, mice: 55
training, primates: 115–134, 157–158
training, rats: 53–55
treats, pigs: 19–21
treats, primates: 131–132
trust relationship with animals: 115–137,
156–157, 177–180
ulcerative dermatitis, mice: 45
vervet monkeys: 97–98, 113–114, 132–133
water as enrichment, primates: 94–96
wheel-running: 49–52
window, cats: 1
wood shavings, mice: 41
wood shavings, primates: 76–77